PROTEIN BIOCHEMISTRY, SYNTHESIS, STRUCTURE AND CELLULAR FUNCTIONS

LAMININS

STRUCTURE, BIOLOGICAL ACTIVITY AND ROLE IN DISEASE

PROTEIN BIOCHEMISTRY, SYNTHESIS, STRUCTURE AND CELLULAR FUNCTIONS

Additional books in this series can be found on Nova's website under the Series tab.

Additional e-books in this series can be found on Nova's website under the e-book tab.

PROTEIN SCIENCE AND ENGINEERING

Additional books in this series can be found on Nova's website under the Series tab.

Additional e-books in this series can be found on Nova's website under the e-book tab.

PROTEIN BIOCHEMISTRY, SYNTHESIS, STRUCTURE AND CELLULAR FUNCTIONS

LAMININS

STRUCTURE, BIOLOGICAL ACTIVITY AND ROLE IN DISEASE

DEMETRIUS C. ADAMS
AND
EMIL O. GARCIA
EDITORS

New York

For permission to use material from this book please contact us:
Telephone 631-231-7269; Fax 631-231-8175
Web Site: http://www.novapublishers.com

Additional color graphics may be available in the e-book version of this book.

Library of Congress Cataloging-in-Publication Data

ISBN: 978-1-62618-568-5

Library of Congress Control Number: 2013935919

Published by Nova Science Publishers, Inc. † New York

Contents

Preface		**vii**
Chapter 1	Laminins: Key Roles from Embryos to Adult Skeletal Muscles *Michèle Beaudry, Valérie Bello and Thierry Darribère*	**1**
Chapter 2	Laminin-332 and Integrins: Signaling Platform for Cell Adhesion and Migration and Its Regulation by N-glycosylation *Yoshinobu Kariya, Yukiko Kariya and Jianguo Gu*	**29**
Chapter 3	Anosmin-1 and Galectin-1 as Laminin Interactors in Nervous System Development and Pathology *Silvia Velasco and Fernando de Castro*	**53**
Chapter 4	A Potential Role of Laminins and 37 kDa Laminin Receptor Precursor (LRP)/67 kDa Laminin Receptor (LR) in Disease *Jisuk Yun*	**69**
Chapter 5	Loss of Laminin Alpha Two *Mahasweta Girgenrath and Tanya Girgenrath*	**87**
Chapter 6	Roles of Laminins in Epithelial Morphogenesis in the Liver *Naoki Tanimizu, Yamato Kikkawa and Toshihiro Mitaka*	**105**
Chapter 7	Role of Laminin on Uterine Remodeling and Embryo Development during Pregnancy *Bruno Zavan, Andréa Mollica do Amarante-Paffaro and Valdemar Antonio Paffaro Junior*	**117**
Chapter 8	Human Gingival Laminin-332 Degradation by Oral Bacteria *Pirjo Pärnänen and Timo Sorsa*	**125**
Chapter 9	Commentary: Laminins as a Modulator of the Axonal Response to Guidance Cues *Tomoyuki Masuda, Takashi Shiga, Hiroyuki Yaginuma and Shuichi Ueda*	**131**
Index		**137**

Preface

Laminins are major proteins in the basal lamina (one of the layers of the basement membrane), a protein network foundation for most cells and organs. Laminins are an important and biologically active part of the basal lamina, influencing cell differentiation, migration, and adhesion, as well as phenotype and survival. In this book, the authors present current research in the study of the structure, biological activity and role in disease of laminins. Topics include the key roles of laminins from embryos to adult skeletal muscle; Laminin-332 and integrins; anosmin-1 and galectin-1 as laminin interactors in nervous system development and pathology; loss of laminin alpha two; role of laminin in epithelial morphogenesis in the liver; laminins in uterine remodeling and embryo development during pregnancy; human gingival laminin-332 degradation by oral bacteria; and laminins as a modulator of the axonal response to guidance cues.

Chapter 1 - There is growing awareness that extracellular matrix components have a central role in driving development, in influencing the outcome of many tissues and in diseases.

However, the authors currently have an incomplete understanding of how these molecules interact with each other *in vivo* in the organization and establishment of the basement membranes. Laminins are extracellular ligands for transmembrane receptors of cell surfaces such as integrins and dystroglycan.

The diversity of laminins can confer to basement membranes specific structures and properties and can generate various signals that develop between the basement membranes, cells and tissues.

Thus, they transmit mechanical forces to cells and/or induce cell signalling events, which are both involved in embryonic development, in adult tissue homeostasis and diseases. Mutations and experimental protein depletion analyses during embryonic development and in adult have provided evidence of the versatile role of laminins in nematodes, flies, frogs, birds and mammals. The contribution of laminins to developmental processes during muscle formation, in adult muscle and muscular diseases is the focus of this review.

Chapter 2 - Laminin-332 (formerly known as laminin-5) is a major component of basement membranes in the skin and other stratified squamous epithelial tissues. Laminin-332 has attracted much attention from both physicians and researchers as a disease-related and functional molecule. The study of human genetic disorders and a knockout mouse model has provided the evidence that laminin-332 is indispensable for adhesive structures in the skin. In wound healing and cancer invasion, overexpression of laminin-332 accelerates cell migration by activating PI3K, PKC, and MAPK signaling pathways. These functional activities of laminin-332 are through interactions with several cell surface receptors including integrins $\alpha3\beta1$ and $\alpha6\beta4$, and syndecans. In addition, recent studies revealed that posttranslational modifications, such as proteolytic processing and *N*-glycosylation, of laminin-332 and of integrins $\alpha3\beta1$ and $\alpha6\beta4$ are important for the laminin-332 activities. This review summarizes the molecular mechanisms and significance of cellular adhesion and migration through laminin-332 as well as integrins $\alpha3\beta1$ and $\alpha6\beta4$. Understanding of these regulatory mechanisms could lead to development of new molecular targeted therapies, particularly in oncology.

Chapter 3 - The laminin family is one of the most studied adhesion molecules. Present in multiple tissues during development and adulthood, in physiological and pathological conditions, laminins are capital for cell adhesion and consequently have been shown to play roles in cell survival, migration and differentiation. Particularly in the nervous system, during development laminin is fundamental for the adhesion of neurons, astrocytes and oligodendrocytes, and as consequence, takes part in the control of the migration of all these cell types as well as in the outgrowth of axons. In the last years, different molecules have been identified as interacting with laminin and two of them, the glycoprotein anosmin-1 and the β-galactoside-binding protein galectin-1, have been shown to be involved in neuronal, astroglial, oligodendroglial development as well as in different neurological diseases. In the present work the authors describe the mechanisms used by anosmin-1 and galectin-1 to bind laminin, and they review the relative relevance of these in the bulk of biological roles that both proteins have been demonstrated to play during nervous system development and in pathology.

Chapter 4 - Laminins, a family of glycoproteins composed of α, ß, and γ chain subunits, present in the basement membrane (BM) of various tissues. Although 16 distinct isoforms have been identified in mammals, their functional differences are not fully understood. Each chain of laminins contains rod-like, globular and coiled regions. Laminin α chain conserves functional sites and chain-specific functional sites.

Laminins play an important role in the organization and structural integrity of BM, and these functions are mediated by various cell surface molecules, such as 37 kDa laminin receptor precursor (LRP)/67 kDa laminin receptor (LR), integrins, dystroglycan, syndecans and lutheran.

LR/LRP is a multifunctional protein which acts as a receptor for laminin and also acts as a receptor for many components; carbohydrates, elastin, green tea polyphenol epigallocatechin-3-gallate (EGCG), various viruses and bacteria and prion proteins. LRP/LR can be a factor in pathological processes. In cancer, LRR/LR is upregulated and its overexpression is associated with tumor cell metastasis and attack. EGCG, a component in green tea, represents anti-allergic, anti-tumor and anti-obesity actions via binding to LRP/LR. The functional domain of LRP mediates the anti-cancer action at the physiological concentration of EGCG (0.1-1 μM). In prion disease, LRP/LR acts as a receptor for (abnormal infectious isoform (PrP^{Sc}) as well as a receptor or co-receptor for cellular prion protein (PrP^{C}) and is essential for the propagation and accumulation of PrP^{Sc} in scrapie-infected cells. LRP/LR has also considered as an alternative therapeutic tools in cancer, prion disease and Alzheimer's disease.

Chapter 5 - Merosin deficient congenital muscular dystrophy type1 (MDC1A) results from mutations in the gene *LAMA2* that encodes for the alpha 2 chain of an indispensible extracellular matrix (ECM) protein known as laminin 211 (also know as merosin). MDC1A is the second most prevalent form of the numerous congenital muscular dystrophies that have been identified. In many cases of this very aggressive disease, premature death of affected children occurs due to either respiratory complications or failure to thrive. While there is no therapy that effectively treats this lethal form of muscular dystrophy, significant progress has been made towards understanding the molecular and biochemical causes for MDC1A and several therapeutic strategies have shown promise in ameliorating the disease pathology in different mouse models of MDC1A. One obvious treatment approach would be to simply replace the missing protein. Indeed, overexpression of ECM proteins such as laminin alpha 1 and agrin significantly improves the pathology of MDC1A since this resolves the primary signaling defect. However, treatment strategies that rely on protein replacement are both challenging to translate into therapy and are often only effective against a singular disease. A more practical and productive approach is to treat some of the secondary manifestations, such as inflammation, necrosis, apoptosis, autophagy, fibrosis and inefficient regeneration of damaged muscle fibers. Not only are these targets more manageable given our current technologies, but they are also common to a host of neuromuscular degenerative diseases, thereby multiplying the potential benefits of an effective treatment. In this review, the authors will present the diverse strategies that are used to treat the primary defect as well as those that focus on the secondary manifestations of the disease. The authors will also discuss the limits of single-mode therapies that target a single disregulated cellular process such as apoptosis or regeneration. Though they may lead to measurable amelioration, a less than optimal outcome is achieved. MDC1A, like many other dystrophies, has multiple disease drivers. Thus, the authors can appreciate that a combinatorial strategy that aims to target more than one pathophysiological process may be more beneficial in tackling the complexities of MDC1A pathology. Indeed there is emerging evidence that such strategies result in more comprehensive recovery and therefore could hold promise for MDC1A and other forms of muscular dystrophy.

Chapter 6 - The liver performs a wide range of metabolic functions, produces serum proteins, and protects the body from infection. To perform these physiological functions efficiently, epithelial cells in the liver, namely hepatocytes and cholangiocytes, establish three-dimensional (3D) tissue structures consisting of hepatic cords and bile ducts. It has been demonstrated that extracellular matrix proteins, especially laminins, play important roles in differentiation of liver stem/progenitor cells and in the formation of tissue structures. Liver progenitor cells acquire metabolic functions similar to hepatocytes and form bile canalicular-like structures *in vitro* in the presence of Matrigel, which contains α1-containing laminin. In 3D culture, liver progenitor cells differentiate along the cholangiocyte lineage and form cysts with the central lumen, a process that depends on α1-containing laminin.

Recently, the authors identified that two laminin isoforms have distinct roles during bile duct development; α1-containing laminin is favorable for inducing cholangiocyte lineage at the initial step of bile duct development, whereas α5-containing laminin is essential for the completion of ductular structure formation. Thus, liver epithelial cells sequentially utilize α1- and α5-containing laminins to drive distinctive processes during bile duct morphogenesis. In this review, the authors describe the roles of laminins in the morphogenesis of hepatocytes and cholangiocytes, in the maintenance of liver stem/progenitor cells, and in pathogenesis.

Chapter 7 - The Laminin – which is made up of three polypeptide chains, α, β and y, joined by bisulfide bonds – interacts with entactin and heparin sulfate to assist the cells in binding to Extracellular Matrix, and therefore participate in cell proliferation and differentiation process. Pregnancy is a phenomenon that, to be viable, depends on major events including the implantation process, trophoblast cells invasion, as well as placental and embryo development. For that reason, the uterus undergoes major morphological changes during pregnancy and depends on the laminin support to ensure the gestation progress. In the non-pregnant mouse uterus, the laminin can be found in the endometrium region exclusively as a component of surface epithelium, glands and capillaries basal membranes, whereas the endometrial stroma is negative. After successful mating the blastocyst apposition and adhesion to the uterine luminal epithelium are assisted by many molecules including laminin. Subsequently, the decidual cells start to synthesize laminin, as well as others molecules, and it is deposited in the endometrial Extracellular Matrix at the time of trophoblast invasion. The integrin repertoire expressed by trophoblasts cells indicate that the laminin and fibronectin support trophoblast invasion until it reaches the maternal spiral arteries. The proper invasion of trophoblast cells is critical to ensure the correct development of the placenta and its protection from maternal immune system rejection, and consequently, it's important to guarantee nutrients, gases, and maternal protective antibodies perfusion into the fetal circulation, producing hormones that control pregnancy, and contributing to the maintenance of a healthy gestation. The role laminin in pregnancy is not restricted to the implantation process, trophoblast invasion and placental development. The major role of this molecule is to support the cellular remodeling that occurs in embryo development process. The absence of any laminin polypeptide chain (α, β and y) is sufficient to cause multiple embryonic defects including syndactyly, failure of neural tube closure, renal agenesis/failure of glomerulogenesis, placental vessel malformations, defects of the neuromuscular junction/glomerulus, skeletal muscle/peripheral nerve, and microvasculature/peripheral nerve. These sets of data demonstrate that laminin is imperative for both uterine remodeling and embryo development during gestation.

Chapter 8 - In this study laminin-332 (LM-332), labelled with ^{35}S- methionine, was isolated from an immortal human oral gingival epithelial cell line. The LM-332 degradation by common oral periodontal bacteria and *Lactobacilli* was detected by SDS-PAGE/ fluorography.

The degradation assays were performed at two different pHs by using bacterial whole cells and 10x concentrated growth media fractions. *Porphyromonas gingivalis* cells at pH 6 and 10x concentrated growth media at pH 6 and 7.3 caused complete and efficient degradation of LM-332 chains. Degradation was also observed as appearance of 30-40 kDa bands by *Prevotella intermedia* cell fraction and *Prevotella nigrescens* 10 x growth media fractions at pH 6. No marked degradation of LM-332 by either fraction or pH occurred by *Lactobacilli*, *Actinobacillus actinomycetemcomitans*, *Fusobacterium nucleatum*, *Peptostreptococcus micros* or *Treponema denticola* cells. Degradation of the basement membrane causes the loss of integrity of the epithelial layer, and this may form an invasion route for oral pathogens, especially through the oral gingival junctional epithelium.

Chapter 9 - Laminins are extracellular matrix molecules known to be abundant in and around the trajectories of developing axons. In addition to their axonal growth-promoting activity, the authors showed that the YIGSR peptide of the lamininβ1 (LAMB1) chain modulates the sensory axonal response to chemoattractive cues derived from the embryonic dorsal spinal cord. They further revealed that such modulationis mediated via the 67-kDa receptor for LAMB1 on sensory axons. By employing culture stripe or growth cone turning assays, the two other groups also provided evidence showing that the responses of developing axons to the repellent Slit2 or attractant Netrin-1 are modulated by the laminin peptide YIGSR. Furthermore, those groups also showed that the modulation of cGMP or cAMP levels in the growth cones of developing axons mimics the modulation of the axonal response to the guidance cues. These findings strongly support the idea that extracellular matrix protein laminins have the potential to modulate the response of neuronal growth cones to the chemotrophic guidance molecules by regulatingthe levels of intracellular cyclic nucleotides.

In: Laminins: Structure, Biological Activity and Role in Disease ISBN: 978-1-62618-568-5
Editors: Demetrius C. Adams and Emil O. Garcia

Chapter 1

Laminins: Key Roles from Embryos to Adult Skeletal Muscles

***Michèle Beaudry[1*], Valérie Bello[2*]* **
and Thierry Darribère[2]
[1]Université Paris 13 Sorbonne Paris Cité, UFR-SMBH,
Laboratoire Réponses Cellulaires et Fonctionnelles à l'Hypoxie. Bobigny France
[2]UPMC Sorbonne Universités (Université Pierre et Marie Curie), UMR CNRS 7622
Laboratoire de Biologie du Développement. Paris, France

Abstract

There is growing awareness that extracellular matrix components have a central role in driving development, in influencing the outcome of many tissues and in diseases.

However, we currently have an incomplete understanding of how these molecules interact with each other *in vivo* in the organization and establishment of the basement membranes. Laminins are extracellular ligands for transmembrane receptors of cell surfaces such as integrins and dystroglycan.

The diversity of laminins can confer to basement membranes specific structures and properties and can generate various signals that develop between the basement membranes, cells and tissues.

Thus, they transmit mechanical forces to cells and/or induce cell signalling events, which are both involved in embryonic development, in adult tissue homeostasis and diseases. Mutations and experimental protein depletion analyses during embryonic development and in adult have provided evidence of the versatile role of laminins in nematodes, flies, frogs, birds and mammals. The contribution of laminins to developmental processes during muscle formation, in adult muscle and muscular diseases is the focus of this review.

Keywords: Skeletal muscles, basement membranes, laminins, development, regeneration, muscular diseases

---

[*] Michèle Beaudry and Valérie Bello contributed equally to this chapter.

Introduction

The natural environment of animal and human cells is constituted by a set of proteins, organized to form extracellular matrices (ECMs). For a long time, the extracellular matrices were seen as stable structures supporting the morphogenesis of tissues and organs.

Now, it seems clear that ECMs are surprisingly dynamic and versatile and that they influence, through direct or indirect means, all steps of embryonic and adult life.

The ECMs are evolutionarily ancient structures, which probably appeared when the first communities of cells have emerged. They are found in organisms ranging from Hydra to Human (Ashkenas et al., 1996). ECMs are composed of heterogeneous meshwork of fibrillar and non-fibrillar components biochemically distinct with different biochemical and physical properties including collagens, different laminins, fibronectin, nidogen, proteoglycans and polysaccharides. These ECM components are secreted and assembled locally into an organized network present in invertebrate, vertebrate embryos and adults. In embryos, the ECM is set up differently during cleavage steps and very generally it is well established before gastrulation, a crucial time in the development of embryos during which the basic body plan is established. ECMs are important for morphogenesis, and mutations in ECM proteins often lead to embryonic lethality. In adults, there are several types of ECMs depending on the molecules that constitute and their organization. In particular, basement membranes (BMs) are thin sheets of highly specialized extracellular matrix on which the various epithelial layers of the body are organized.

They also surround single cells such as skeletal muscle fibres, adipocytes, and peripheral nerves. At the ultrastructural level, BMs possess two layers: the basal lamina and the reticular lamina. The basal lamina is formed by two layers, a clear layer called the lamina lucida and a dense layer called the lamina densa.

According to high-power TEM micrographs, BMs form ultrathin sheets ranging from 100 nm to 10 μm thick. Components of the BMs are multi-domain proteins of high molecular weight that emerged about 500 millions years ago during evolution (Hynes, 2012). They either polymerize or bind to other proteins suggesting that BMs are composed of a two-dimensional scaffold of collagen IV and a network of polymerized laminins (Yurchenco and Patton, 2009). Nidogen has been proposed to generate connections between the two polymers (Fox et al., 1991). BMs are important to maintain the organization and homeostasis of tissues since from worms to human mutations that affect components of BMs lead to severe phenotypes including early embryonic death, muscular dystrophies, kidney and cardiovascular defects and eye, ear and brain malformations.

Within BMs, the reticular lamina contains collagens I, III, and V and various proteoglycans. The lamina densa of basal lamina contains type IV collagen fibres, entactin/nidogen, perlecan, and heparan sulfate proteoglycans. The lamina lucida of basal lamina is largely composed of collagen IV, perlecan, nidogen and laminins. Laminins represent the most important non-collagenous components of the basal lamina. In collaboration with the network developed by collagen molecules, laminins participate to mechanical properties of the basal lamina. Laminins also undertake interactions with other extracellular proteins and with cell surface receptors. Through these connections, laminins have a central role in the formation, organization and physiology of the basal lamina (Aumailley et al., 2005). The ECMs containing laminins are dynamic structures of cell environments, whose composition and spatial organization differ between species, developmental steps, tissues and obviously with age in long-lived humans. However in most species, it provides support and anchorage for cells, segregates tissues from one another. ECMs also sequester and store growth factors, present growth factors to their receptors, sense and transduce mechanical signals (Rozario and DeSimone, 2010).

For several years, various model organisms have been used to study ECMs structures and functions and in particular the role played by laminins. These include nematode worms, fruit flies, zebrafish, frogs, chicken and mice. They all contributed in various ways and significantly to our understanding of the laminin functions during embryogenesis, in adults and during postnatal diseases. In this review, we will focus on laminin functions in skeletal muscles from embryos to adults.

Laminin Discovery, Characterization and Expression

Laminins, the most abundant non-collagenous proteins in BMs, were initially purified from the transplantable Engelbreth-Holm-Swarm (EHS) mouse sarcoma and from mouse parietal yolk sac carcinoma. Laminins are large glycoproteins (400 to 900 kDa) composed by assembly of three polypeptide chains α, β and γ genetically distinct and interconnected by disulfide bridges. Eleven genes encode for the eleven chains of the laminin family. To date, five α-chains, four β-chains and three γ-chains have been characterized in mouse and human (Miner and Yurchenco, 2004). Their various combinations lead to the existence of at least 16 different trimmers from mouse and human, if the major splice variants and a partially described variant are included (Yurchenco, 2011).

The α-chain averages 400 kDa in size and 160 nm in length. The 1000 C-terminal amino acid residues contribute to five homologous 150–180 amino acid globular motifs, which are known as the G domain (LG 1-5 domains, figure 1). The G domains are the major site for laminin interactions with cells and are also found in many other ECM proteins (for a review on laminin G domains see Durbeej, 2010; Aumailley, 2013). Both-β- and γ-chains average 200 kDa, with short arms of 60 and 40 nm in length respectively, and they lack a G domain.

Laminin isoforms are heterotrimeric chains assembled intracellularly and then secreted into the ECMs. Posttranslational maturation differs between different laminins, during development and within tissues leading to variable glycosylation and proteolytic processing. The chains assemble into a cross with two or three asymmetric short arms and one long arm (Figure 1). Some of the α- and β-chains have truncated short arms leading to a T-shaped heterotrimer. The C-termini of α-, β-, and γ-chains form the core of the laminin cross.

The core is a triple α-helical coiled-coil domain of 77-nm in length. It extends by the G domain of α-chains involved in interactions with cells (Figure 1). The three short arms are the N-termini of the chains and exhibit alternative rod-like and globular domains (Figure 1). These globular domains mediate the laminin polymer formation of BMs. Interestingly, differential expression of laminin chains have been shown with respect to the stage of development and tissue distribution, presumably reflecting diverse *in vivo* functions.

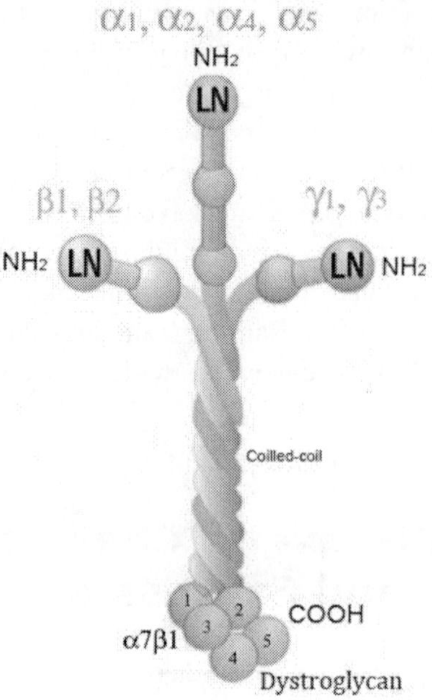

Figure 1. Schematic representation of the structure of laminins. Rotatory shadowing electron microscopy revealed that laminins have a cross-shaped form with one long and three short arms. The various isoforms are characterized by the presence of laminin α-chains combined with laminin β- and γ-chains. The short arms of all the chains have two globular domains, which are separated by rod-like domains. The central long arm of all laminins is a coiled-coiled structure composed of all three chains. In skeletal muscle cells, the globular domains (LG 1-3) of the COOH region of α-chains interact with α7β1 integrins. The dystroglycan cellular receptor binds the LG 4-5 domains.

One of the difficulties in describing laminin distributions and expressions reside in the fact that the detection methods must targeted the three chains. These difficulties have often led to an imperfect description of the chain composition of laminins assembled in BMs. Anyway, through combinations of the methods of biochemistry, molecular biology, immunohistochemistry, today the various chains can be fairly accurately located (Table 1). Laminin-111 is the first laminin expressed during mammalian development since mRNAs for the β1- and γ1-chains are detected at the 2–4 cell stage in mice and since the α1-protein is observed in the morula (Cooper and McQueen, 1983; Dziadek and Timpl, 1985). Through development and in the adult, laminin-111 assembles to form a network associated with those formed by collagen IV, they form the backdrop for the most basal laminae.

As the α1-chain, the α5-chain is present in the BMs of skeletal muscle cells (Sorokin et al., 1997). In these cells, the α5-chain expression is down regulated with muscle differentiation and in the adult it is primarily localized at the neuromuscular junction. The α2-chain is primarily a mesenchymal laminin subunit found in the BMs surrounding skeletal muscles (Paulsson et al., 1991). The α4-subunit is particularly highly expressed by mesenchymal cell types such as skeletal muscle cells (Liu and Mayne, 1996). The β2-chain is specifically present at the neuromuscular junction while the β1-chain is expressed in the rest of the sarcolemmal BMs (Hunter et al., 1989). Finally, the γ1-chain is the most ubiquitous laminin chain found in all BMs and thus surrounds skeletal muscles. Its presence is required for BMs formation.

Table 1. Major expression sites of laminin chains in mice or human

Chains	Tissue distribution	Laminin variants
α1	Early embryo, around somites, placenta, kidney glomeruli, renal tubules, mammary glands and saliva, ovaries, uterus, testis, retina, intestine, liver, lung, pancreas	α1-β1-γ1 (laminin-111) α1-β2-γ1 (laminin-121)
α2	Trophoblast, somites, skeletal and cardiac muscles, peripheral nerves, intestine, brain	α2-β1-γ1 (laminin-211) α2-β2-γ1 (laminin-221) α2-β1-γ3 (laminin-213)
α3	Skin, stratified epithelia	α3-β1-γ1 (laminin-311) α3-β2-γ1 (laminin-321) α3-β3-γ2 (laminin-332)
α4	Primarily mesenchymal cells, fetal muscles, adult muscles, lung, nerve, blood vessels, bone marrow	α4-β1-γ1 (laminin-411) α4-β2-γ1 (laminin-421) α4-β2-γ3 (laminin-423)
α5	Epithelia, muscle progenitors, kidney, synapse blood vessels	α5-β1-γ1 (laminin-511) α5-β2-γ1 (laminin-521)
β1	Ubiquitous	α1-β1-γ1 (laminin-111) α2-β1-γ1 (laminin-211) α2-β1-γ3 (laminin-213) α3-β1-γ1 (laminin-311) α4-β1-γ1 (laminin-411) α5-β1-γ1 (laminin-511)
β2	Neuromuscular junction, glomerulus	α1-β2-γ1 (laminin-121) α2-β2-γ1 (laminin-221) α3-β2-γ1 (laminin-321) α4-β2-γ1 (laminin-421) α4-β2-γ3 (laminin-423) α5-β2-γ1 (laminin-521) α5-β2-γ2 (laminin-522) α5-β2-γ3 (laminin-523)
β3	Skin and epithelia	α3-β3-γ2 (laminin-332)
γ1	Most tissues	α1-β1-γ1 (laminin-111) α1-β2-γ1 (laminin-121) α2-β1-γ1 (laminin-211) α2-β2-γ1 (laminin-221) α3-β1-γ1 (laminin-311) α3-β2-γ1 (laminin-321) α4-β1-γ1 (laminin-411) α4-β2-γ1 (laminin-421) α5-β1-γ1 (laminin-511) α5-β2-γ1 (laminin-521)
γ2	skin and epithelia	α3-β3-γ2 (laminin-332) α5-β2-γ2 (laminin-522)
γ3	Out of basement membranes in epithelia, nerve, brain	α2-β1-γ3 (laminin-213) α4-β2-γ3 (laminin-423) α5-β2-γ3 (laminin-523)

In 1990, laminin-211 (α2-β1-γ1), named merosin, has been described as the most abundant laminin isoform in the BMs of adult skeletal muscles (Ehrig et al., 1990). At the end of the long arm of laminin-211, there are five globular repeats (LG domains) that protrude and bear binding sites for its cell receptors, dystroglycan and integrins (Timpl et al., 2000). However, as mentioned above, other laminin isoforms are present during myogenesis at the neuromuscular and the myotendinous junctions.

Laminin Receptors

There are two major types of cell surface receptors for laminins, the proteins of the integrin family and the dystroglycan. Among the integrin family, nine have been described to interact with laminins including α1β1, α2β1, α5β1, α3β1, α6β1, α6β4, αvβ3, αvβ5, and α7β1. The latter is known to be the receptor for laminin in muscle cells (Burkin et al., 1999). The α7β1 integrin is abundant in differentiated muscle fibres and is present in adult skeletal muscles. The extracellular domain of α7β1 integrin binds within laminin domains G1-G3. The intracytoplasmic domain of α7β1 integrin can link to the actin filaments of skeletal muscle fibres via talin and vinculin. Once bound to laminin, the intracellular cytoplasmic domain can also activate focal adhesion kinases (FAK), small rho GTPases, and MAPK pathways to regulate cellular activities (Hynes 2002).

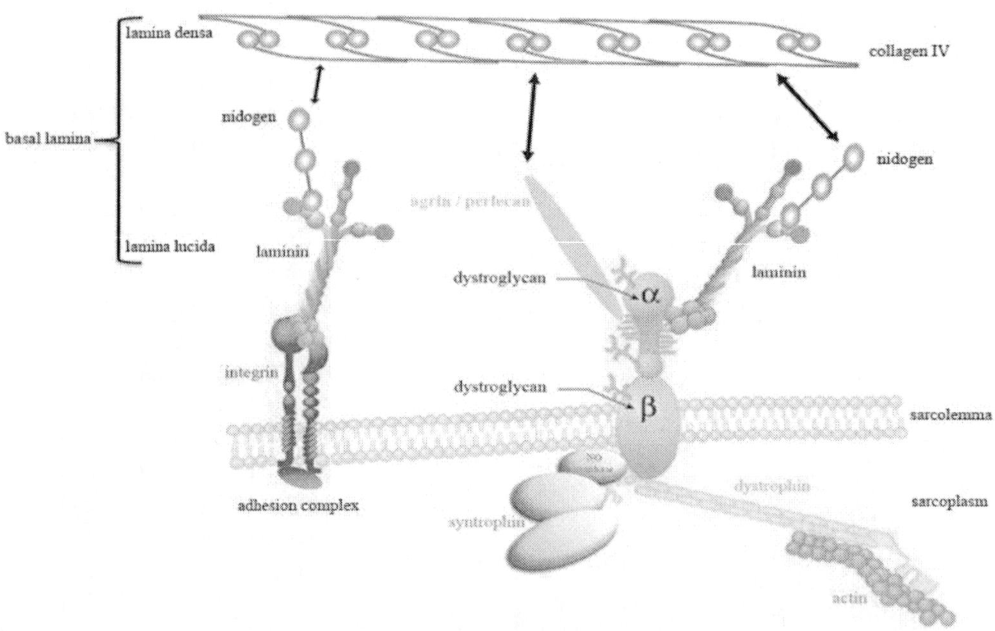

Figure 2. Molecular structure of the basement membrane of skeletal muscles. At the cell surface of skeletal muscles, BMs are composed of a two-dimensional scaffold of collagen IV and a network of polymerized laminins. The laminins are anchored to the cell surface by interactions of the long arms of α-chains with α7β1integrin and α-dystroglycan cellular receptors. Laminins polymerize by interactions between their short arms. The laminin and collagen networks are linked by nidogen and heparan sulfates (←→).

Dystroglycan is a large glycoprotein, which is present at the cell surface as two noncovalently linked subunits, α- and β-dystroglycan. The subunits are produced by post-translational modifications of a precursor protein, transcribed from a single gene (Ibraghimov-Beskrovnaya et al., 1993). The β-dystroglycan is a transmembrane glycoprotein and the α-dystroglycan an extracellular glycoprotein, that acts in ligand binding with agrin, perlecan, biglycan, α1- and α2-chains (Ervasti and Campbell, 1993). In skeletal muscles, dystroglycan binding to laminins occurs among the mucinous O-linked carbohydrate chains located in the mid-neck region of α-dystroglycan. The cytoplasmic tail of β-dystroglycan interacts with the actin filaments of skeletal muscle fibres via the dystophin/utrophin/syntrophin complex (Figure 2). It seems increasingly clear that α7β1 acts as an anchor linking muscle cells to the tendon, whereas dystroglycan is involved in maintaining integrity of the BMs of muscle fibres. Thus BMs containing laminins provide platforms for multiple signalling events.

The other types of laminin receptors correspond to non-integrin proteins. One distinguishes the 32/67kDa laminin receptor, the 67kDa elastin-laminin receptor, the galactoside-binding lectin, the galactosyltransferase, the heparin sulfate proteoglycans and Ig-related basal cell adhesion molecule.

Laminins in Invertebrate Muscle Formation

In the free-living, transparent *Caenorhabditis elegans* nematode, the body musculature is formed by 95 cells localized just beneath the specialized epidermis that secretes the cuticular exoskeleton (White et al., 1976). The body wall muscles exert their force on the cuticle to generate locomotion. Interposed between the muscle cells and the cuticle, there is a BM containing laminins. *Caenorhabditis elegans* has four laminin genes *epi-1, lam-3, lam-1* and *lam-2* encoding four different chains. The *epi-1* gene encodes αB- chain that is similar to mammalian α5-chain, the *lam-3* gene encodes the αA-chain that is similar to mammalian α1- and α2-chain, the *lam-1* gene encodes the β-chain and the *lam-2* gene encodes the γ-chain, suggesting two isoforms that differ by their α-subunit assembly. Both forms have overlapping but distinct localization patterns. Both forms are detected near the end of gastrulation. The αA-chain is found around pharynx muscle cells and intestinal precursors and along the muscle cells. The αB-chain is concentrated on epidermis and along muscle cells (Huang et al., 2003). In larval and adult animals, the only chain detected in body wall muscle and accessory muscles is the αB-chain. Mutations within each laminin α-subunit cause cell-cell adhesion defects, mispositioned adhesion complexes, disorganized cytoskeleton and abnormal ECM. On the surface of body wall muscle cells, laminin organises into a polygonal array whereas in mutants, muscle cells fail to adhere to the overlying epidermis. Muscle adhesion proteins and myofibrils are mispositioned and in the epidermis the cytoskeleton is defective at the point of attachment of myofibres. The *epi-1* mutants exhibit disruptions of the body wall muscles and failures of muscle adhesion to epidermis. Cell polarity is disturbed and sarcomeres are disorganized. The studies in *Caenorhabditis elegans* show that laminin molecules assemble to form an initial network where muscle cells can anchor. Then, the primitive network allows the assembly of the BM over the muscle tissues.

In *Drosophila melanogaster*, somatic muscles form the external muscular wall of the larva, they are similar to skeletal muscles of vertebrates. Each of these muscles is constituted of a single mononuclear fibre unlike vertebrate muscle, which consists of several fibres. Each segment has a muscle pattern consisting in thirty different muscles. This pattern varies between segments along the antero-posterior axis (Bate 1990). All mature muscles share common physiological and contractile properties. Muscles attach to specialized cells called tendon cells and both bind to the underling ECM containing laminins. The *Drosophila* genome encodes four laminin chains: two α-chains (α1,2 and α3,5) similar to vertebrate α3 and α5, one β-chain and one γ-chain. These chains form two trimers, laminin A (α3,5; β1; γ1) and laminin W (α1,2; β1; γ1) (Fessler et al., 1987; Kusche-Gullberg et al., 1992). The second laminin α chain, α1,2 (encoded by *wing blister, wb*), is most similar to vertebrate α1- and α2-chains. They are distributed in BMs, including those of somatic and visceral muscles. A null mutation of the α3,5 chain gene causes embryonic defects of somatic muscles (Yarnitzky and Volk, 1995). The defects concern the attachment of the ventral oblique muscles to the ectoderm and the stabilization of the shape of myotubes during their extension and contraction (Yarnitzky and Volk 1995). The β-chain distribution resembles that of α-chains and is enriched at muscle attachment sites. In the absence of the β-chain, no trimers could be made, defects in the attachment of ventral oblique muscles are observed as for laminin α1,2 mutants. In addition, it causes detachment of other muscles from tendon cells and muscles adopt a spherical shape (Urbano et al., 2009). So, in β-chain mutants, α3,5 and α1,2 chains do not localize at muscle attachment sites suggesting that, around muscles, the accumulation of the α3,5 and α1,2 chains in the BMs requires the presence of β-chains. Furthermore, BMs are absent and the remaining extracellular material appears disorganized and diffuse. However in the absence of the β-chain, the assembly of matrix components into BM surrounding muscle attachment sites does not seem to be affected, even though laminin-α1,2-β1-γ1 is a component of this matrix (Urbano et al., 2009). Together, the data suggest that the complete absence of laminins leads to abnormal accumulation of major BM components resulting in the complete absence of BMs and subsequently prevented muscle development (Urbano et al., 2009).

The spectrum of these phenotypes in invertebrates suggests that laminins have crucial functions throughout invertebrate muscle development.

Laminins in Vertebrate Muscle Formation

In vertebrate embryos, most skeletal muscles, except some head muscles, derive from somites. The somites form transient structures in pairs in rostro-caudal direction by epithelial condensations and segmentation of paraxial mesoderm on both sides of the neural tube and notochord. The first segmented somites appear in the anterior part of the embryo and the number varies from one species to another. Although all vertebrates form somites, most works have carried on chick, zebrafish and amphibian embryos with some additional information from the inactivation of genes performed by technical knockout in mice.

The paraxial mesoderm is a loose mesenchymal tissue composed of groups of unorganized cells (Pourquié, 2001). The conversion to individualized somites goes through condensation of cells to form a cohesive polarized epithelium. This transformation involves changes in adhesion between cells and between cells and the ECM containing laminins and fibronectin. The regulation of these molecules appears to be controlled by the expression of the transcription factor Paraxis expressed early during epithelialization.

Somite segmentation requires the expression of genes in a cyclic pattern regulated by a negative feedback loop in a time dependent manner, known as "clock and wave front" model, a conserved patterning mechanism across species (Cooke and Zeeman, 1976). Spatiotemporal somite formation (somitogenesis) requires expression of genes of the Notch and Wnt pathways and morphogen gradients of Wnt, FGF, and retinoic acid.

In the caudal region of the paraxial mesoderm, FGF and Wnt pathways maintain cells to a mesenchymal state (Aulehla and Pourquié, 2010). They control a periodic activity of the Notch pathway, which controls downstream cyclic expression of genes involved in the generation of somites (Hofmann et al., 2004). When cells leave the caudal region, an increasing level of retinoic acid establishes the somite polarity and stops the cyclic expression of genes. Then the somites develop distinct dorso–ventral compartments (Figure 3).

After the epithelialization and the segmentation of somites, cells of the ventral part of somites undergo an epithelial to mesenchymal transition leading to generate sclerotome progenitor cells. A set of genes (Pax1, Pax9, Scleraxis) specifies their commitment to a chondrogenic fate. These progenitors give rise to vertebrae and ribs (Figures 3 and 4).

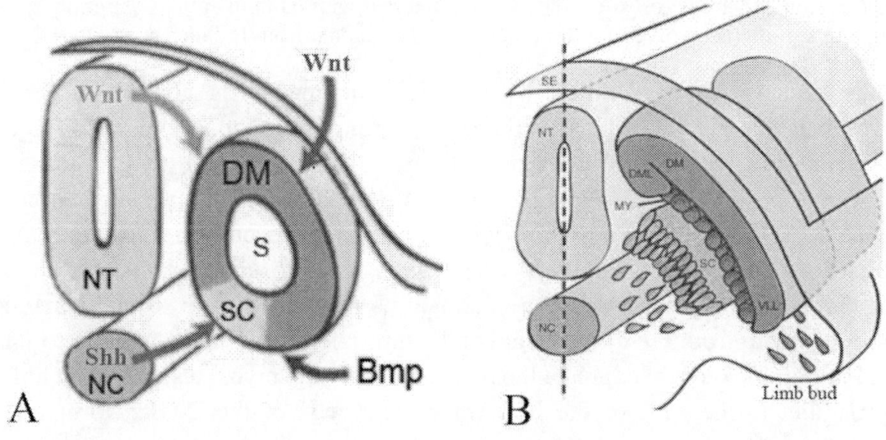

Figure 3. The embryonic origin of trunk and limb skeletal muscles. (A) The somites form from paraxial mesoderm on both sides of the neural tube and notochord. At early somitogenesis, somites are spherical structures with a central cavity, the somitocoele (S). They form the epithelial somites. Specification of somitic lineages occurs in response to signals from adjacent structures. The signalling molecules Wnts, sonic hedgehog (Shh) and Bmp specify the somite to form the sclerotome (SC) and dermomyotome (DM). Wnts and Bmp maintain the somites in an undifferentiated state and Shh induces the formation of the sclerotome. These signals inhibit or promote the determination of somite cells in muscle precursors. (B) During somitogenesis, the sclerotome cells segregate. At the dorsal medial lip (DML) of the dermomyotome, cells delaminate and populate the myotome to produce epaxial muscle progenitors. At the ventral lateral lip (VLL), cells delaminate and populate the myotome. They are the source of progenitors for hypaxial muscles. At the level of the limb buds, cells delaminate from the VLLs to give rise to limb muscles. (Adapted from Bentzinger et al., 2012).

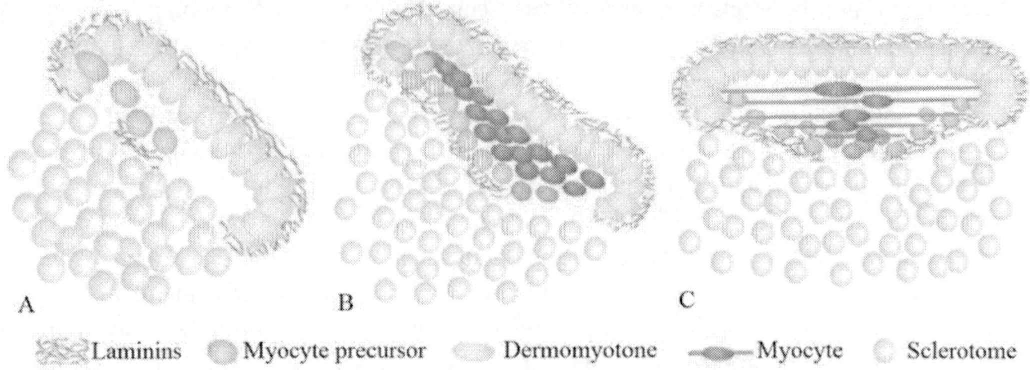

Figure 4. A potential model for myotomal basement membrane formation. (A) Transverse representation. A layer of thick ECM, containing laminin 111, which connects it to the adjacent structures, surrounds each somite. The dorsomedial lip cells delaminate and initiate the deposition of laminin-111 to form the myotomal ECM. Dorso-medial lip cells synthesize α5-chain, which begin to form laminin-511 incorporated in the extracellular matrix with collagen IV, perlecan and nidogen to form a continuous, stable myotomal BM that separates the myotome and sclerotome. (B) Transverse representation. Myocyte precursors populate the myotome in a laminin dependant guidance. Few cells stay in close contact with laminin at the interface between myotome and sclerotome. (C) Longitudinal representation. In the median region of the myotome-sclerotome interface, the BM is discontinuous. The myocytes elongate in the rostro-caudal axis and interact with laminin at their extremities suggesting that laminin is involved for their elongation. (Adapted from Bajanca et al., 2006).

On the dorsal part of somites, a sheet of columnar cells appears and forms the dermomyotomes that give rise to dermatomes that form dermis and myotomes. Myotomes are the source of cell progenitors for trunk, limbs, diaphragm, and tongue muscles. The myotomes form by involution of cells from dermomyotomes (Figure 4). It then consists of mononucleated cells called myocytes. Myotome formation occurs in two waves. First, the cells of the dorsomedial lip delaminate, orient themselves rostro-caudally, differentiate into myocytes and cover the underside of the dermatome (Figure 4). The second wave concerns cells that delaminate from the dorsomedial and ventrolateral lips. They migrate to the rostral and caudal region of somites, populate the myotome, become myocytes and then elongate in a direction parallel to the cells of the first wave. The cells of the DMLs, form the epaxial myotomes. They differentiate rapidly to form the deep muscles of the back, associated with the spine. The cells of the VLLs give rise to the hypaxial myotomes, which form somitic buds that invade the lateral mesoderm and give rise to the intercostal muscles, the muscles of the abdominal wall, the superficial muscles of the back, the diaphragm and the muscles of the tip of the tongue. At the level of the limb bud, cells from the lateral edge of the dermomyotomes delaminate and migrate into the lateral mesoderm, where they differentiate to form limb and limb girdle muscles (Figure 3). A cascade of myogenic regulatory factors (MRFs), including MyoD, Myf5, Myf6 and myogenin, are temporally and sequentially expressed to initiate all these events and myotome cell differentiation into skeletal muscle fibres. Prior to differentiation, MyoD and Myf5 are expressed in proliferating cells, while myogenin and Myf6 are expressed in myoblasts that have initiate their differentiation (for exhaustive reviews see Francetic and Li, 2011; Bentzinger et al., 2012). The formation of somites in zebrafish, xenopus, chicken and mouse depends on particular interactions between cells of the mesoderm and ECM molecules. Around each immature somite, laminin-111 is probably the first constituent of the extracellular matrix since the gene encoding the laminin α1-chain, *Lama1*, is transcribed in the presomitic mesoderm in the mouse and the protein deposited around somites in avian (Duband et al., 1987; Anderson et al., 2009). Laminin α1-chain is synthesized under the control of Shh signalling and assembles with β1- and γ1-chains into a laminin-111 network (Figure 3). While entering the myotome during somite differentiation, muscle progenitors begin to synthesize laminin α5-chain, which presumably begin to assemble with β1- and γ1-chains. To form the myotomal BMs, that separate the myotome and sclerotome, laminin-511 is progressively incorporated into the extracellular network established previously by laminin-111, at least in mice (Figure 4).

All skeletal muscles are formed from proliferating myoblasts that differentiate, fuse into multinucleate myofibrils, assemble a contractile apparatus, and form synapses with motor neurons. Myogenesis has different levels of complexity in organisms ranging from jellyfish to humans. To understand the mechanisms involved in myogenesis, the use and characterization of cellular and animal models were needed. Thus, the development of muscles has been particularly studied *in vivo* in model species such as drosophila, zebrafish, xenopus, chicken and mouse. Current knowledge on myogenesis have also benefited from the contribution of mutations in animal models. Thus, strains of mice have been developed in order to provide models of muscular dystrophy studies. These animals have physiological defects that mimic those observed in human patients.

During the course of life, there are at least three distinct myogenic programs: embryonic, fetal and adult myogenesis. The latter will be discussed in another section below. Very interestingly, a small number of muscle stem cells ensures the establishment of functional adult muscles composed of millions of differentiated cells. This is probably due to the extraordinary capacity for self-renewal of these cells during development and in the adults. This could also be due to a fine regulation of cell fate choice by components of the ECMs.

During embryonic development, it is widely accepted that there are two waves of myoblast fusion known as primary and secondary myogenesis resulting in individual muscles that are composed of multiple myofibres (Figure 5). A first wave of proliferation and migration of precursors is observed forming embryonic myoblasts that will be the first to fuse leading to primary myofibres. It is followed by the migration of a second group of myoblasts, called fetal myoblasts, which surround the primary myofibres (Vivarelli et al., 1988). They fuse in turn, with each other or with the primary myofibres, to give secondary myofibres. With the progression of secondary myogenesis, fetal myoblasts interact and fuse with the ends of both primary and secondary myofibres. During myogenesis, myofibres express different isoforms of myosin heavy chains leading to generate future fast- and slow-contracting muscles of the adult. The primary myotubes will form predominantly slow fibres of muscles constituting posture muscles. The secondary myotubes provide fast fibres involved in the movement. Later in development, another type of muscle precursor cells, the adult satellite cells, are present between laminins of the basal lamina and the myofibres (Figure 5). Myoblasts, derived from these satellite cells, can generate cells that are responsible for adult muscle growth and regeneration. In the postnatal period, the muscle growth depends on the fusion of myoblasts to the ends of myofibres, resulting in increased muscle length and girth.

During the two myogenic phases, BMs are continuously remodelled and differential expression of laminin chains is observed. In mice, during the primary myogenesis at embryonic day 11, laminin-211 and laminin-511 are the major laminins of the BMs. At embryonic day 11.5, the laminin α1-chain is still present, but restricted to the ends of myofibres. During the secondary myogenesis, expression of the laminin α4-chain increases at embryonic day 14. By embryonic day 15, it is deposited around and throughout the secondary myofibres in the BMs. At the seventh week of human gestation, the α2-chain is detected and its maximum expression appears around the twenty-first week of gestation. The α4-chain is expressed at the sixteenth week of gestation (Petäjäniemi et al., 2002). During myogenesis in humans, the α5-chain is the major laminin α-chain. The α1-chain is detected during the development of myotendinous junction (MTJ).

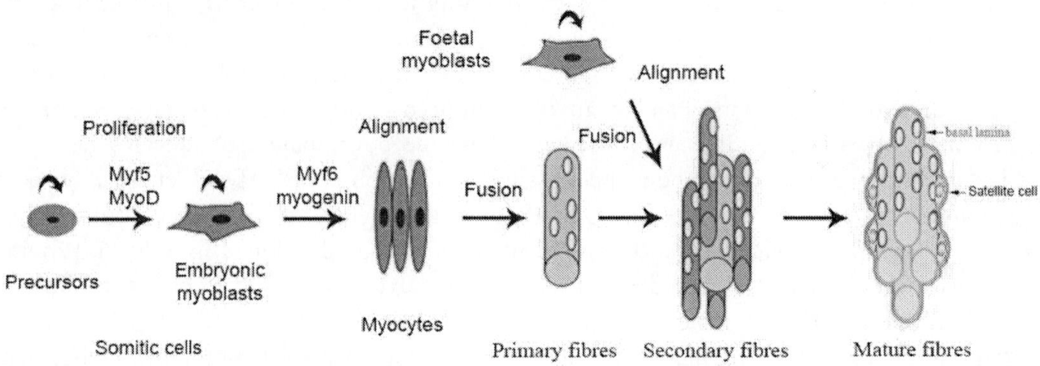

Figure 5. Formation of skeletal muscle fibres. Four myogenic regulatory factors (MRFs) have been identified in vertebrates: Myf5, MyoD, myogenin and Myf6. MyoD and Myf5 are involved in the early stages of myogenesis in the determination of myogenic precursors, whereas myogenin and Myf6 are required for the differentiation and maintenance of the differentiated phenotypes. Muscle precursors and myoblasts proliferate, and then they leave irreversibly the cell cycle and become postmitotic. They align with each other and fuse to form myotubes precursors of primary muscle fibres also called embryonic fibres. A second wave of fusion occurs around the second third of fetal life to form secondary fibres. They are derived from fetal myoblasts that align and fuse around the primary fibres also called fetal fibres. Around mature fibres and under the basal lamina, myoblasts form satellite cells of adult muscles.

Perinatally both in human and mouse, the levels of laminin α4- and α5-subunits decrease at birth and at the end of the first postnatal week, they are not detectable at the sarcolemma.

In rat skeletal muscle cultures, laminin substrate promotes myogenesis by enhancing proliferation of myoblasts, by stimulating their motility, and by favouring the bipolar shape characteristic of fusing cells (Foster et al., 1987). When primary mouse skeletal myoblasts are plated onto laminin, they differentiate into postmitotic cells and are able to fuse (von der Mark and Ocalan, 1989). Cell attachment assays, performed with the established mouse C2C12 myogenic cell line and primary myoblasts, show that they adhere to laminin-111 and -211 isoforms and more rapidly on laminin-211 (Schuler and Sorokin, 1995). In C2C12 cells, merosin expression is upregulated, while laminin-111 expression is downregulated and both proteins promote myoblast fusion and myotube formation. Interestingly, only merosin is involved in the stability and the survival of myotubes by preventing apoptosis (Vachon et al., 1996). Taken together, these data suggest that laminins-111 and -211 play distinct roles in myogenesis at least *in vitro*. *In vivo*, their precise functions in myogenesis are imperfectly known. From mouse models of inactivation of genes for β2, α2, and α4-chains, the most significant data on skeletal muscles have been obtained with the invalidation of the gene encoding the α2-chain. Unexpectedly myogenesis does not seem to be affected. Myotube formation decreases, although not abolished, and merely smaller muscle fibres are observed at birth (Miyagoe et al., 1997). This is probably linked to a compensation effect between the different chains of laminin during myogenesis. In contrast, laminins that contain the α2-chains seem to be crucial in adult muscles, this point will be discussed in more detail in following sections.

In zebrafish *Danio rerio*, the myotome is composed of the superficial slow muscle fibres and the deep fast muscle fibres (for a review Ochi and Westerfield, 2007). Orthologs to ten of the known mammalian laminin genes were found. The α1-, α2- and α4-chains are expressed in the embryonic and larval myotomes. The α1-chains are transiently present in the somites before the 24 hours post-fertilization stage. The α2-chains are the most highly expressed in muscles throughout development and in fast and slow fibres. The α4-chains are expressed in the anterior and ventral region of somites. During development, its expression becomes restricted to cells located between fibres and cells at the apex of myotomes, a region containing muscle progenitor cells. Later, the α4-chains are detected around cells between fast and slow fibres. The mRNA encoding β1-, and γ1-chains are all detected in the somitic cells during late somitogenesis and subsequently at the level of the trunk muscles of later larval stages (Sztal et al., 2011). In the *sly/lamc1* mutation that disrupts the gene for γ1-chain, BMs are disorganized, non-functional and somites are mis-shaped (Parsons et al., 2002). Zebrafish mutants of α2-chains exhibit progressive degeneration of skeletal muscles and die few days after fertilization (Gupta et al., 2012). Muscles have damaged at the site where the myotomes attach and exhibit progressive detachment of myofibres.

In *Xenopus laevis*, the transcripts for α2-, α3-, α4- and α5-chains are found in the presomitic mesoderm. As the embryos develop, transcripts are localized in somites (Ahmad, 2009). Using antibodies against the EHS laminin, the protein is found around the somites and later at the BM of skeletal muscles (Wedlich et al., 1989). There are only few works on the role of laminin during the formation of skeletal muscle in amphibians. However, indirect data were obtained by performing mutation or depletion of the laminin receptors. A dominant negative form of integrin β1-subunit alters somite formation (Marsden and DeSimone, 2003). Dystroglycan depletion and overexpression of dystroglycan deleted of the cytoplasmic domain result in a dramatic disorganization of the myotome and a misorientation of muscle cell precursors within somites because they are unable to anchor to the intersomitic ECM containing laminins (Hidalgo et al., 2009).

As for invertebrates, these data suggest that the interaction between laminins and their receptors are a prerequisite for muscle cell precursors organization within the somites.

Laminins in Adult Skeletal Muscles

At the ultrastructural level the muscle BMs consist of two layers: the basal lamina and the reticular lamina. The reticular lamina is connected to the endomysium when the basal lamina is directly linked to the sarcolemma. The basal lamina possesses an electron lucent layer, the lamina lucida and an electron-dense layer, the lamina densa (Yurchenco and Patton, 2009). Basal lamina surrounds each muscle fibre and passes through the junctions between muscle and nerves. As we have seen above, during embryonic development of skeletal muscles, the predominant isoform is laminin-111, formed by α1-, β1- and γ1-chains. Laminin α2-chain participates in the formation of laminin-211 formed by α2-, β1- and γ1-chains, and laminin-221 formed by α2-, β2- and γ1-chains. Together, these laminins are major constituents of skeletal muscle basal lamina. The immunogold electron microscopy show that the merosin is present at the lamina lucida of the basal lamina of human skeletal myofibres, connecting the basal lamina to the neighboring plasma membrane of muscle fibres (Shibuya et al., 2003). The loss of these laminins deteriorates the adhesion of muscle fibres, weakens the sarcolemma and increases sensitivity to apoptosis (see next section).

Laminins at the Neuromuscular Junctions

The synapse, at the skeletal neuromuscular junction (NMJ) forms in mammalian embryos and matures later, during the postnatal period and is stable in young adult mice. At the NMJ, pre and postsynaptic structures are precisely differentiated. In the presynaptic region of the motoneuron, an increase in synaptic volume and area, an increase of synaptic vesicles, clustering of vesicles at the active zones, and polarization of the pre-synaptic membrane are observed. In the post-synaptic region of muscle cells, there are an accumulation of mitochondria and neurotransmitter receptors. Laminins act as an organizer of differentiation. They have important roles in the development, function and regeneration of the nervous system, particularly for the development of a totally functional NMJ. Laminins are concentrated in the synaptic cleft as the NMJ develops and are kept to maintain the structure and function of adult NMJ.

The NMJ is a synapse formed between motor nerves and skeletal muscle cells. From the spinal cord, motor neurons emit axons that reach the muscles when myoblasts are aligning and have just started to fuse. At a week of development in mice, synapses begin to form with an accumulation of synaptic vesicles, the appearance of the BM and the formation of active zones. The NMJ use diverse neurotransmitter molecules depending on the species: acetylcholine (Ach) in vertebrates, glutamate in drosophila. Both of them induce muscle contraction (Wu et al., 2010). During the NMJ development, muscle cells secrete the ECM to constitute the basal lamina and the motoneuron also secretes factors such as agrin (Smirnov et al., 2005). In the NMJ formation, the active zones of synapses located at the motoneuron extremity must be adjusted to the muscle endplate to become innervated. The active zone is the synaptic vesicle release site; molecules are known to control active zone density and extracellular matrix including laminins (Clarke et al., 2012). Laminins containing α4-, α5-, β2- and γ1-chains are synthetized by skeletal muscle cells and deposited on the sarcolemmal surface. The basal lamina of the synaptic cleft is enriched in α2-, α4-, α5-, β2-chains at very high level with high α2-β2-γ1 in the mature cleft (Sanes, 2003). Loss of α2-chain is accompanied by a modification in the form and complexity of the structure of the NMJ leading to muscular dystrophy (Smirnov et al., 2005). The lack of α5-chain delays NMJ development but has non-effect on adult NMJ (Samuel et al., 2012). Mice lacking α4-chain are viable and fertile, but lack precise apposition of pre- and postsynaptic structures without compromising the architecture and topology of the synapse (Patton et al., 2001). Mice lacking α4- and α5-chains exhibit presynaptic abnormalities and delayed postsynaptic maturation

(Nishimune et al., 2008). Laminins containing β2-chains are implied in the development of functional NMJ (Noakes et al., 1995). During development, NMJ have both β1- and β2-chains whereas adult NMJ have only β2-chain. In laminin-β2 mutant mice all synaptic laminin heterotrimers are depleted, presynaptic development is affected, the NMJ fails to mature and animals die 2–4 weeks after birth due to a lack of proper neuromuscular transmission. Noakes, et al., (1995) and Patton et al., (1998) demonstrated that the lack of proper neuromuscular transmission is due to Schwann cells that migrate into the synaptic cleft inhibiting neurotransmission to the muscle. Moreover, laminins containing β2-chains are involved in the organization of calcium channels, which are required in neurotransmission (Nishimune et al., 2008; Carlson et al., 2010). The β2-chain seems to binds directly to the P/Q-type calcium channels and clusters them at the active zones (Nishimune et al., 2004). Wu et al., (2010) indicate that mice lacking β2-chain can form a synapse but do not maintain active zones. It can also be shown that the disruption of the β2-chain and calcium channels interactions *in vivo* reduce active zones (Wu, et al., 2010). Pierson syndrome, a recently defined lethal disease in the first postnatal months in human, is caused by mutations in the gene encoding the β2-chain. The hallmarks of Pierson syndrome are congenital nephrotic syndrome accompanied with muscular and neurological developmental defects. *Lamb2–/–* mice are a model for Pierson syndrome; they exhibit defects in the kidney glomerular barrier, in the development and organization of the neuromuscular junction (Miner et al., 2006).

Taken together the results suggest that the myofibres secrete molecules to organize the presynaptic development and that α4-, α5- and β2-chains are directly involved in postsynaptic membrane maturation (Nishimune, et al., 2008).

Laminins and Muscle Regeneration

Skeletal muscle has the ability to regenerate after injury and the capability to adapt to physiological situations such as growth or training. In the adult skeletal muscle, the ability to regenerate depends on satellite cells, a cell population located between the basal lamina and the sarcolemma of their associated muscle fibres (Figure 5). Satellite cells derive from embryonic cells of the somites. They appear in late fetal myogenesis and are localized in the fibres and under the BMs, in an environment called 'the niche" (Figure 5). Recent studies have revealed non-muscle stem cells able to participate in muscle regeneration (Zammit et al., 2006). But most of these cells are found to have an inefficient regenerative potential. It is why today, the satellite cells are defined as the main muscle progenitors in adults. Under physiological conditions, satellite cells are mitotically quiescent, whereas, on injury, they are activated and indispensable for regeneration of skeletal muscles. They integrate signals from the microenvironment that surrounds them, in particular from the basal lamina and its components including laminins. Satellite cells will move from a quiescent to an activated stage. Following activation, satellite cells leave their niche, move outside of the basal lamina and start to divide. Their descendants form adult skeletal myoblasts that undergo multiple rounds of division. Then, they differentiate to fuse and form multinucleated myofibres. A fraction of these myoblasts eventually leave the cell cycle, return to quiescence and become self-renewing cells leading to new satellite cells associated with the regenerated fibres. Their progeny will be able of efficient new regeneration (Montarras et al., 2005).

The major laminin-211 and 221 receptors in skeletal muscle are integrin α7β1 and dystroglycan. As mutations in the α7 subunit of integrin gene cause congenital myopathy, Rooney et al., (2009a) investigated the role of integrin α7β1 in muscle regeneration after cardiotoxin (a snake venom with membrane-damaging activity inducing tissue necrosis) was used to induce muscle damage in a model of α7 integrin–null-mice. The experiments indicated that in this mice model, satellite cells activation is delayed after cardiotoxin injection and muscle exhibited defective regeneration compared to wild type mice.

Satellite cells are skeletal muscle progenitor cells responsible for postnatal growth, repair and regeneration (Chargé and Rudnicki, 2004; Montarras et al., 2005). The regenerative capacity of skeletal muscle depends on a pool of resident satellite cells, located between the sarcolemma and the basal lamina of the myofibre (Mauro, 1961). These quiescent satellite cells are quickly activated in response of muscle damage (injury, exercise, disease), they divide to provide myoblasts for muscle repair (Moss and Leblond 1971; Collins et al., 2005) and are responsible for generating myoblasts in adult skeletal muscles. During this activation these stem cells, which are normally in an undifferentiated quiescent state proliferate and differentiate to fuse with each other or with existing multinucleated myofibres (Schulz and McCormick, 1994; Hawke and Garry, 2001; Chargé and Rudnicki, 2004; Zammit et al., 2006; Rooney et al., 2009a; Relaix and Zammit, 2012). It has been shown that enhanced expression of laminin α7β1 receptor activates satellite cell proliferation and differentiation in a dystrophic skeletal muscle (Rooney et al., 2009b).

As loss of laminin- α2 expression was observed in this model of α7 integrin–null-mice (Rooney 2009a), it is hypothesized that loss of laminin participates in defective muscle regeneration. To explore whether reduced laminin deposition could account for the defective muscle regeneration phenotype observed in the model of α7 integrin–null-mice, laminin-111 was injected in muscle three days before cardiotoxin injection (Rooney et al., 2009a). Laminin-111 is normally found in the extracellular matrix of skeletal mice and human muscles only during embryonic development and is absent in dystrophic skeletal muscles. The injected laminin infiltrated the whole muscle and restored myogenic repair and regeneration in α7 integrin–null-mice suggesting that laminin-111 is involved in muscle regeneration. The mechanism involved in repair and regeneration by laminin-111 injection in muscle damaged by cardiotoxin in this model is not yet understood. The authors concluded that laminin-111 is able to restore capacity of regeneration in a mice model of congenital myopathy and could be a therapeutic agent.

Similarly, Goudenege et al., (2010) demonstrated that intramuscular injection of laminin-111 could increase muscle strength and resistance in Duchenne muscular dystrophy (DMD) in mdx mice, a good model for understanding this pathology in humans. Currently, there is no treatment for DMD. Different strategies for cell and gene therapies are being investigated in animals and humans. Myoblast transplantation (MT) is a potential therapy in the case of DMD. Clinical trials and animal experiments were conducted; the treatment induces the formation of new muscle fibres, new satellite cells, dystrophin positive in mice and the formation of small myofibrils in patients. Recent studies by Silva-Barbosa et al., (2008) indicate a beneficial effect of laminin on the MT.

As previously described above, the injection of laminin-111 in mdx mice (Rooney et al., 2009a) protected muscle against damages induced by exercise, enhanced α7-integrin expression, improved myoblast proliferation and migration *in vitro*. These results suggest that laminin-111 could be a therapeutic agent for DMD cell therapy. However, the results of

Gawlik et al., (2011) are not consistent with those of Rooney et al., (2009a). In this study, the authors explored the ability of laminin-111 using transgenic mice overexpressing α1-chain in the skeletal muscle to prevent DMD. Despite of laminin deposition laminin-111 overexpression did not prevent mice from DMD. Taken all together the results indicate that the therapeutic use of laminin-111 for DMD requires more investigations.

If human myoblasts are injected in anterior tibialis muscles from an immunodeficient mice model subjected to local irradiation, enhanced laminin deposit occurred. A better engraftment of human myoblasts was observed after local irradiation. Moreover, it is observed that the same human myoblasts respond to laminin *in vitro* with increased proliferation, cell survival, migration. These results *in vivo* and *in vitro* suggest that laminin level may be important in cell therapy strategies.

Diseases Involving Laminins and Its Receptors

There are more than 200 neuromuscular diseases, disorders affecting muscle and its nerve control. 80-90% are of genetic origin and the diseases can appear at any age. They can only affect muscles, or muscles and heart or muscles and brain or spinal cord and muscles. The remaining 10-20% are autoimmune diseases such as rheumatism. Among these diseases, some are the result of changes affecting the expression of one or more chains of laminin or abnormalities of their post-translational maturation. Others involved modifications of the cell surface receptors for laminins.

Congenital muscular dystrophies (CMDs) are rare genetic diseases. The term includes several different diseases characterized by muscle damage resulting in muscle weakness, hypotonia and motor difficulties that appear at birth or in first months of life. Skeletal muscles show a dystrophic process characterized by muscle cell degeneration accompanied by the presence of young regenerating cells tending to offset the loss cell due to degeneration. This muscle damage can be complicated by orthopedic deformations, respiratory difficulties and variable clinical manifestations of eyes and central nervous system.

Since the first patient was described in 1903 by Batten, the clinical and molecular classification of CMDs has considerably evolved, so that today one distinguishes three main groups of CMDs based on the defects in structural proteins of skeletal muscle fibres, nucleus, disorders of protein glycosylation, and mutations in selenoprotein N (SEPN1) a protein of the endoplasmic reticulum. The first group where the myofibre proteins are involved includes the CMD with primary merosin deficiency (MDC1A), the CMD linked to collagen VI deficiency (type Ullrich, DMCU) and a form with mutations of the integrin α7-subunit. The second group, called dystroglycanopathies, a CMD linked to mutations affecting the N-glycosylation and O-mannosylation of α-dystroglycan, includes the Walker-Warburg syndrome, the muscle-eye-brain syndrome (MEB), the Fukuyama's disease (FCMD), the early muscular dystrophy with microcephaly and mental retardation. A third group consists of the other forms such as the rigid spine syndrome, with lissencephaly and peripheral neuropathy, with mitochondrial abnormalities, with mental retardation and congenital cataract, with primitive cardiomyopathy and finally, severe form caused by sporadic mutations in the gene encoding lamins A/C of the nucleus.

Merosin, collagen IV, α7 integrin and α-dystroglycan participate to maintain a connection between inside and outside of muscle cells. This anchorage system between the two environments allows the muscle cells to adapt to mechanical stress, especially when they undergo deformations during muscle contractions. In CMD, the deficit in one or the other of these proteins will cause a dysfunction of the anchorage system. The link between the exterior and interior of fibres would be broken, leading, over time, to an alteration of the cell membranes by repeated contractions and, ultimately, a weakening of the whole muscles. With advances in genetics and molecular diagnostic techniques, the genetic basis of this very heterogeneous group of congenital muscular dystrophies are much better known. In particular, in order to investigate the molecular mechanisms involved in the congenital muscular dystrophies and to test new therapeutic approaches, animal models that mimic the genetic abnormalities and clinical signs of the diseases have been developed. Currently for the dystroglycanopathies affecting the glycosyltransferase genes, there is a mouse model of the MEB syndrome (genetic defect in the POMGnT1 gene), a mouse model for the MDC1D (genetic defect in the LARGE gene) and a mouse model of the Fukuyama's disease (genetic abnormality in the fukutin gene). There are also mouse models of the Ullrich congenital muscular dystrophy and the MDC1A syndrome with primary merosin deficiency.

Merosin-deficient congenital muscular dystrophy (MDC1A) is the most common CMD and is caused by the absence or reduction of α2-chain in skeletal muscles. MDC1A represents 50% of CMD cases and has an incidence of 1 in 21500, nearly half of the cases in European countries caused by recessive mutations in the *LAMA2* gene (Holmberg and Durbeej, 2013). The *LAMA2* gene of mouse is mapped to chromosome 10A4-B1. In human, the gene is localized on chromosome 6 at 6q22-q23 and it spans 260 kb containing 64 exons. It encodes the α2-chain of the main laminin isoforms expressed by myofibres and peripheral nerves. Actually, more than 800 mutations in the gene have been identified. They concern both coding and non-coding sequences. MDC1A is characterized by skeletal muscle weakness, and hypotonia. It is also often accompanied by brain abnormalities characterized by white matter defects appearing at two year old. Also, focal cortical dysplasia and epilepsy have been observed in patients (Messina et al., 2010). Most mutations in the *LAMA2* gene cause complete deficiency in α2-chain but there are also cases of partial deficiency (Nissinen et al., 1996). It was shown that most severe cases were observed in complete absence of laminin α2-chain.

In 1955, a spontaneous mouse model of muscular dystrophy (MD), the *dy* mouse, showing severe progressive muscular dystrophy, and a death between 2 and 6 months of age has been described (Michelson et al., 1955). This model has allowed the identification of a mutation in the chromosome 10, at the location of the *LAMA2* gene, showing a deficient expression of merosin in the peripheral nerve, skeletal and cardiac muscles (Sunada et al., 1994). Also, homozygous mice for the dy^{2J} allele of *LAMA2* gene have been extensively studied showing a deletion in the aminoterminal domain of α2-chain providing an animal model for study the molecular basis of congenital muscular dystrophy (Xu et al., 1994a, b). This strain exhibits progressive necrosis of muscle fibres that starts around 10 days of age. This phenotype is followed by variation in fibre size, loss of muscle fibres, ineffective muscle regeneration, leading to accumulation of fibrosis, similar to the pathology of muscular dystrophy in humans (Vainzof et al., 2008). Since, mouse models with targeted disruption of *LAMA2* gene have been generated leading to various severities of phenotypes and expression of α2-chain. Analyses and comparison of the different phenotypes of these different mouse models have helped in the characterization of MDC1A and in the development of different therapeutic approaches. They all show that mutants display growth retardation and die from more or less severe muscular dystrophies after birth. Biochemical, genetical, immunohistochemical and ultrastructural analyses often show the presence of the other BM components while an almost complete absence of intact BMs is found around the muscle fibres. They also show that mutations in α2-chain result in defective laminin polymerization or interactions between laminin and its cell receptors within the skeletal muscle BMs.

Mutations in the dystrophin protein complex as the integrin α7-subunit lead to a MD. Mutations in the dystrophin gene caused DMD. These types of mutations sometimes causes disruptions of the sarcolemma which results in the entry of extracellular matrix components especially calcium in the muscle fibres. It was demonstrated in a model of mdx mice that calcium entry caused hypercontractions (Claflin and Brooks, 2008). However, it seemed that the degree of sarcolemma damage did not correlate with the severity of the disease (Gawlik et al., 2010). Recent studies on a zebrafish model suggest that the severity of the disease caused by α2-chain mutations should be related to myofibres detachment and not because of sarcolemma disruptions (Hall et al., 2007).

It was further noted that in the case of patients with MDC1A and mice presenting laminin-211 deficiency, the dystrophin-glycoprotein complex was identical to healthy controls (Tomé et al., 1994; Miyagoe et al., 1997). Peeling of isolated muscle fibres showed that the ϒ-actin involved in costamers was stably linked to the dystrophin. On the contrary, this was not the case in mdx mice, which exhibited a significant decrease in all components of the dystrophin-glycoprotein complex excluding laminin-211 where no stable link between ϒ-actin and dystrophin was observed (Rybakova et al., 2000). Altogether, these results suggest that laminin-211 is essential to the integrity of muscle and loss of α2-chain leads to very severe myopathy.

However, compensatory effects may occur. In the case of the loss of the α2-chain, an increase of α4-chain is observed (laminin-411); but α4-chain, which does not have N-terminal domain, cannot link to the cells and therefore does not replace the loss of the α2-chain (Talts et al., 2000). Other studies of Moll et al. (2001) show that the mini-agrin, a chick construct that comprised the amino-terminal agrin domain (Denzer et al., 1998), binds to laminin-411 and dystroglycan and is sufficient to amend muscle function in laminin α2-deficient mice. Myopathy, which results in mice, is less severe (Moll et al., 2001). So the expression of mini-agrin by systemic delivery of recombinant virus strongly ameliorates the muscular phenotype of laminin-alpha2-deficient mice (Qiao et al., 2005).

Moreover, it has been shown (Rooney et al., 2009a, b; Goudenege et al., 2010) that overexpression of laminin-111 reduced MD in case of α2 chain-deficient mice. Recently, using Engelbreth-Holm-Swarm-derived mouse laminin-111 systemically delivered to the muscles of laminin-α2-deficient mice, it has been shown that this treatment prevents muscle pathology, improves muscle strength, and increases life expectancy (Rooney et al., 2012). The treatment also prevents apoptosis in primary human MDC1A myogenic cells establishing the potential replacement therapies for the treatment of MDC1A (Rooney et al., 2012). Although these genetic therapies in mouse are interesting, the application of this approach into clinical treatments seems difficult. To overcome these difficulties, several pharmacological options have been tested, which is expected to provide other clinical treatments. The objectives of these options are to inhibit apoptosis, proteasomal and autophagy degradation of proteins in dystrophic mice. For example, halofuginone, an analogue of a plant alkaloid, prevents progression of fibrosis, resulting in an amelioration of the dystrophic phenotype (Nevo et al., 2010). An analogue derivative of losartan, an anti-hypertensive drug, given to three-week-old mice restores muscle regeneration following toxin-induced injury (Elbaz et al., 2012). Also, it improves the weight, strength and locomotory activities in the mouse model for laminin-α2-deficient congenital muscular dystrophy (Meinen et al., 2012).

Further works are needed to understand the complex microenvironments created mainly by BM components such as laminins, their satellite cells and progenitors during homeostasis, diseases and repair of skeletal muscles. But we can currently consider that there are at least five targets that could be studied as the main axes of potential therapeutic approaches. The first target could be to strengthen the adhesion between muscle cells and the BMs. This approach could increase the level of expression of adhesion proteins, such as α7β1 integrins and dystroglycan that can reinforce the link between the laminins and the actin network. The second target could be to promote the integrity of sarcolemma by consolidating and increasing proteins involved in membrane integrity of the muscle fibres. It is also possible to stimulate the compensation effect by protein counterparts to recover the membrane proteins associated with anchorage and connection to the network of ECMs proteins at the muscle fibres. The fourth target could be to foster the cohesion and of the extracellular matrix proteins to allow an efficient organization of these proteins in the cell extracellular compartment of the muscle fibres. Finally, it would be interesting to better understand the interactions between satellites cells and the ECM proteins of the niche in order to find the better way to stimulate their proliferation and differentiation and thus to promote the regeneration of skeletal muscle fibres

Conclusion

The extracellular matrix and BMs are the non-cellular component present within all tissues and organs. They provide essential physical scaffolding for the cellular constituents and initiate crucial biochemical and biomechanical cues that are required for skeletal muscle morphogenesis, differentiation and homeostasis. Numerous studies on the structure and function of laminins have been completed and it is very clear today that laminins play an extremely important role during the development and regeneration of skeletal muscle and in pathologies. It stands out from the studies that laminin-111, present only during muscle development and -211 present in adulthood have a pivotal role in muscle function. In mice, most knockouts of laminin chains are lethal, such as γ1-chain, due to the lack of BM formation. If the mice survive, they develop diseases, according to the subunits and their tissue distributions. The loss of the α2-chains cause congenital muscular dystrophy (MDCIA1) for which there is currently no treatment. Moreover, loss of α2-chains could have deleterious effects on satellite cells whose location is between the BMs and the sarcolemma, which could affect muscle regeneration cycles since the activation of satellite cells is involved in these phenomena. The β2-chain (concentrated at skeletal muscle synapses) deficiency, known in humans as Pierson syndrome, results in disorganized and defective neuromuscular junctions and is lethal in the first postnatal months. These various studies indicate that chain mutations or loss of one of the laminin subunits cause deleterious or lethal effects. Although laminin-111 is expressed very early in the blastocyst and only during the embryonic period, the injection of laminin-111 appears to have a beneficial effect in DMD as described by Rooney et al., (2009a, b) and Goudenege et al. (2012). Indeed, in this case, the laminin-111 can partially restore the lost link between the sarcolemma and the extracellular matrix. These results suggest that in the future, DMD and other muscular dystrophies could be improved, stopped or even treated by injection of a recombinant laminin-111 allowing the muscle to be protected during muscle contraction. Moreover, Goudenege et al. 2010 showed that the prior injection of laminin-111 in cell therapy by myoblast transplantation improves the transfer. Taken together these results provide a complementary view of therapeutic action of laminin-111 in muscular dystrophies.

Finally, despite numerous studies and therapeutic approaches, currently there are no treatments for these disorders induced by muscle laminin deficiencies. However, current studies on the interactions between the sarcolemma and laminins of the extracellular matrix components are very important in the future development of treatments for muscular dystrophies.

Acknowledgments

Many scientists have made significant contributions to the laminin field. The authors apologize to colleagues whose work they were unable to cite because of space limitation.

References

Ahmad, N. (2009). Temporal and Spatial Expression Pattern of Four Laminin Alpha Chains in Xenopus laevis. *J. Biol. Sciences,* 9, 128-136.

Anderson, C., Thorsteinsdóttir, S., and Borycki, A.G. (2009). Sonic hedgehog-dependent synthesis of laminin α1 controls basement membrane assembly in the myotome. *Development,* 136, 3495–3504.

Ashkenas, J., Muschler, J., and Bissell, M.J. (1996). The extracellular matrix in epithelial biology: shared molecules and common themes in distant phyla. *Dev. Biol.,* 180, 433–444.

Aulehla, A. and Pourquié, O. (2010). Signaling gradients during paraxial mesoderm development. *Cold Spring Harb Perspect Biol.,* 2, a000869.

Aumailley, M., Bruckner-Tuderman, L., Carter, W.G., Deutzmann, R., Edgar, D., et al. (2005). A simplified laminin nomenclature. *Matrix Biol.,* 24, 326-32.

Aumailley, M. (2013). The laminin family. Cell Adh Migr. In press.

Bajanca, F., Luz, M., Raymond, K., Martins, G.G., Sonnenberg, A., Tajbakhsh, S, Buckingham, M., and Thorsteinsdóttir, S. (2006). Integrin alpha6beta1-laminin interactions regulate early myotome formation in the mouse embryo. *Development,* 133, 1635-1644.

Bate, M. (1990). "The embryonic development of larval muscles in Drosophila." *Development,* 110, 791-804.

Bentzinger, C.F., Wang, Y.X., and Rudnicki, M.A. (2012). Building Muscle: Molecular Regulation of Myogenesis. *Cold Spring Harb Perspect. Biol.,* 4, pii: a008342.

Batten, F. (1903). Three cases of myopathy, infantile type. Brain 26, 147–148.

Burkin, D.J., and Kaufman, S.J. (1999). The alpha7beta1 integrin in muscle development and disease. *Cell Tissue Res.,* 96, 183–190.

Carlson, S.S., Valdez, G., and Sanes, J.R. (2010). Presynaptic calcium channels and α3-integrins are complexed with synaptic cleft laminins, cytoskeletal elements and active zone components. *J. Neurochem.,* 115, 654-66.

Chargé, S.B., and Rudnicki M.A. (2004). Cellular and molecular regulation of muscle regeneration. *Physiol. Rev.,* 84, 209-238.

Claflin, D.R., and Brooks, S.V. (2008). Direct observation of failing fibres in muscles of dystrophic mice provides mechanistic insight into muscular dystrophy. *Am. J. Physiol. Cell Physiol.,* 294, 651-658.

Clarke, G.L., Chen, J., and Nishimune, H. (2012). Presynaptic Active Zone Density during Development and Synaptic Plasticity. *Front. Mol. Neurosci.,* 5, 12.

Collins, C.A., Olsen, I., Zammit, P.S., Heslop, L., and Petrie, A. (2005) Stem cell function, self-renewal, and behavioral heterogeneity of cells from the adult muscle satellite cell niche. *Cell,* 122, 289–301.

Cooke, J., and Zeeman, E.C. (1976). A clock and wavefront model for control of the number of repeated structures during animal morphogenesis. *J. Theor. Biol.,* 58, 455-476.

Cooper, A.R., and MacQueen, H.A. (1983). Subunits of laminin are differentially synthesized in mouse eggs and early embryos. *Dev. Biol.,* 96, 467–71.

Denzer, A. J., Schulthess, T., Fauser, C., Schumacher, B., Kammerer, R.A., Engel, J. and Ruegg, M.A. (1998). Electron microscopic structure of agrin and mapping of its binding site in laminin-1. *EMBO J.,* 17, 335-343.

Duband, J.L., Dufour, S., Hatta, K., Takeichi, M., Edelman, G.M., and Thiery, J.P., (1987). Adhesion molecules during somitogenesis in the avian embryo. *J. Cell Biol.,* 104, 1361–1374.

Durbeej, M. (2010). Laminins. *Cell Tissue Res.,* 339, 259-268.

Dziadek, M., and Timpl, R. (1985). Expression of nidogen and laminin in basement membranes during mouse embryogenesis and in teratocarcinoma cells. *Dev. Biol.,* 111, 372–382.

Ehrig, K., Leivo, I., Argraves, W.S., Ruoslahti, E., and Engvall, E. (1990). Merosin, a tissue-specific basement membrane protein, is a laminin-like protein. *Proc. Natl. Acad. Sci. USA,* 87, 3264-3268.

Elbaz, M., Yanay, N., Aga-Mizrachi, S., Brunschwig, Z., Kassis, I., Ettinger, K., Barak, V., and Nevo, Y. (2012). Losartan, a therapeutic candidate in congenital muscular dystrophy: studies in the dy(2J) /dy(2J) mouse. *Ann. Neurol.,* 71, 699-708.

Ervasti, J.M., and Campbell, K.P. (1993). A role for the dystrophin-glycoprotein complex as a transmembrane linker between laminin and actin. *J. Cell Biol.,* 122, 809-823.

Fessler, L.I., Campbell, A.G., Duncan, K.G., and Fessler, J.H. (1987). Drosophila laminin: characterization and localization. *J. Cell. Biol.,* 105, 2383–2391.

Foster, R.F., Thompson, J.M., and Kaufman, S.J. (1987). A laminin substrate promotes myogenesis in rat skeletal muscle cultures: analysis of replication and development using antidesmin and anti-BrdUrd monoclonal antibodies. *Dev. Biol.,* 122, 11-20.

Fox, J.W., Mayer, U., Nischt, R., Aumailley, M., Reinhardt, D., Wiedemann, H., Mann, K., Timpl; R., Krieg, T., and Engel, J. (1991). Recombinant nidogen consists of three globular domains and mediates binding of laminin to collagen type IV. *EMBO J.,* 10, 3137-3146.

Francetic, T., and Li, Q. (2011). Skeletal myogenesis and Myf5 activation. *Transcription,* 2, 109-114.

Gawlik, K.I., Akerlund, M., Carmignac, V., Elamaa, H., and Durbeej, M. (2010). Distinct roles for laminin globulardomains in laminin alpha1 chain mediated rescue of murine laminin alpha2 chain deficiency. *PLoS One,* 5, e11549.

Gawlik, K.I., Oliveira, B.M., and Durbeej, M. (2011). Transgenic expression of Laminin α1 chain does not prevent muscle disease in the mdx mouse model for Duchenne muscular dystrophy. *Am. J. Pathol.,* 178, 1728-1737.

Goudenege, S., Lamarre, Y., Rousseau, J., Frenette, J., Skuk, D. and Tremblay, J.P. (2010). Laminin-111: *A Potential Therapeutic Agent for Duchenne Muscular Dystrophy Mol. Ther.,* 18, 2155–2163.

Goudenege, S., Lebel, C., Huot, N.B., Dufour, C., Fujii, I., Gekas, J., Rousseau, J., and Tremblay, J.P. (2012). Myoblasts derived from normal hESCs and dystrophic hiPSCs efficiently fuse with existing muscle fibers following transplantation. *Mol. Ther.,* 20, 2153-2167.

Gupta, V.A., Kawahara, G., Myers, J.A., Chen, A.T., Hall, T.E., Manzini, M.C., Currie, P.D., Zhou, Y., Zon, L.I., Kunkel, L.M., and Beggs, A.H. (2012). A splice site mutation in laminin-α2 results in a severe muscular dystrophy and growth abnormalities in zebrafish. *PLoS One,* 7, e43794.

Hall, T.E., Bryson-Richardson, R.J., Berger, S., Jacoby, A.S., Cole, N.J., Hollway, G.E., Berger, J., and Currie, P.D. (2007). The zebrafish candyfloss mutant implicates extracellular matrix adhesion failure in laminin alpha2-deficient congenital muscular dystrophy. *Proc. Natl. Acad. Sci. USA,* 104, 7092-7097.

Hawke., T.J., and Garry, D.J. (2001). Myogenic satellite cells: physiology to molecular biology. *J. Appl. Physiol.,* 91, 534–551.

Hidalgo, M., Sirour, C., Bello, V., Moreau, N., Beaudry, M., and Darribère, T. (2009). In vivo analyzes of dystroglycan function during somitogenesis in Xenopus laevis. *Dev. Dyn.,* 238, 1332-1345.

Hofmann, M., Schuster-Gossler, K., Watabe-Rudolph, M., Aulehla, A., Herrmann, B.G., and Gossler, A. (2004). WNT signaling, in synergy with T/TBX6, controls Notch signaling by regulating Dll1 expression in the presomitic mesoderm of mouse embryos. *Genes Dev.,* 18, 2712-2717.

Holmberg, J., and Durbeej, M. (2013) Laminin-211 in skeletal muscle function. *Cell Adhesion and Migration,* 7, 1–11.

Huang, C.C., Hall, D.H., Hedgecock, E.M., Kao, G., Karantza, V., Vogel, B.E., Hutter, H., Chisholm, A.D., Yurchenco, P.D., and Wadsworth, W.G. (2003). Laminin {alpha} subunits and their role in C. elegans development. *Development,* 130, 3343–3358.

Hunter, D.D., Shah, V., Merlie, J.P., and Sanes, J.R. (1989). A laminin-like adhesive protein concentrated in the synaptic cleft of the neuromuscular junction. *Nature,* 338, 229–234.

Hynes, R.O. (2002). Integrins: bidirectional, allosteric signaling machines. *Cell,* 110, 673-87.

Hynes, R.O. (2012). The evolution of metazoan extracellular matrix. *J. Cell Biol.,* 196, 671-679.

Ibraghimov-Beskrovnaya, O., Milatovich, A., Ozcelik, T., Yang, B., Koepnick, K., Francke, U., and Campbell, K.P.(1993). Human dystroglycan: skeletal muscle cDNA, genomic structure, origin of tissue specific isoforms and chromosomal localization. *Hum. Mol. Genet.,* 2, 1651-1657.

Kusche-Gullberg, M., Garrison, K., MacKrell, A.J., Fessler, L.I., and Fessler, J.H. (1992). Laminin A chain: expression during Drosophila development and genomic sequence. *EMBO J.,* 11, 4519–4527.

Liu, J., and Mayne, R. (1996). The complete cDNA coding sequence and tissue-specific expression of the mouse laminin α4 chain. *MatrixBiol.,* 15, 433–437.

Marsden, M., and DeSimone, D.W. (2003). Integrin-ECM interactions regulate cadherin-dependent cell adhesion and are required for convergent extension in Xenopus. *Curr. Biol.,* 13, 1182-1191.

Mauro, A. (1961) Satellite cell of skeletal muscle fibres. *J. Biophys. Biochem. Cytol.,* 9, 493–495.

Messina, S., Bruno, C., Moroni, I., Pegoraro, E., D'Amico, A., Biancheri, R., Berardinelli, A., Boffi, P., Cassandrini, D., Farina, L., Minetti, C., Moggio, M., Mongini, T., Mottarelli, E., Pane, M., Pantaleoni, C., Pichiecchio, A., Pini, A., Ricci, E., Saredi, S., et al., (2010). Congenital muscular dystrophies with cognitive impairment. A population study. *Neurology,* 75, 898–903.

Meinen, S., Lin, S., and Ruegg, M.A. (2012). Angiotensin II type 1 receptor antagonists alleviate muscle pathology in the mouse model for laminin-alpha2-deficient congenital muscular dystrophy (MDC1A). Skelet Muscle. 2, 18.

Michelson, A.M., Russell, E.S., and Harman, P.J. (1955). Dystrophia muscularis: a hereditary primary myopathy in the house mouse. *Proc. Natl. Acad. Sci. USA,* 41, 1079–1084.

Miner, J.H., and Yurchenco, P.D. (2004). Laminin functions in tissue morphogenesis. *Annu. Rev. Cell Dev. Biol.,* 20, 255–284.

Miner, J.H., Go, G., Cunningham, .J, Patton, B.L., and Jarad, G. (2006). Transgenic isolation of skeletal muscle and kidney defects in laminin β2 mutant mice: implications for Pierson syndrome. *Development,* 133, 967-975.

Miyagoe, Y., Hanaoka, K., Nonaka, I., Hayasaka, M., Nabeshima, Y., Arahata, K., Nabeshima, Y., and Takeda, S. (1997). Laminin alpha2 chain null mutant mice by targeted disruption of the Lama2 gene: a new model of merosin (laminin 2)-deficient congenital muscular dystrophy. *FEBS Lett.,* 415, 33-39.

Moll, J., Barzaghi, P., Lin, S., Bezakova, G., Lochmüller, H., Engvall, E., Müller, U., and Ruegg, M.A. (2001). An agrin minigene rescues dystrophic symptoms in a mouse model for congenital muscular dystrophy. *Nature,* 413, 302-307.

Montarras, D., Morgan, J., Collins, C., Relaix, F., Zaffran, S., Cumano, A., Partridge, T., and Buckingham, M. (2005). Direct isolation of satellite cells for skeletal muscle regeneration. *Science,* 304, 2064-2067.

Moss, F.P., and Leblond, C.P. (1971). Satellite cells as the source of nuclei in muscles of growing rats. *Anat. Rec.,* 170, 421–435.

Nevo, Y., Halevy, O., Genin, O., Moshe, I., Turgeman, T., Harel, M., Biton, E., Reif, S., and Pines, M. (2010). Fibrosis inhibition and muscle histopathology improvement in laminin-alpha2-deficient mice. *Muscle Nerve,* 42, 218–229.

Nissinen, M., Helbling-Leclerc, A., Zhang, X., Evangelista, T., Topaloglu, H., and Cruaud, C. (1996). Substitution of aconserved cysteine-996 in a cysteine-rich motif of the laminin alpha2-chain in congenital muscular dystrophy with partial deficiency of the protein. *Am. J. Hum. Genet.,* 58, 1177-1184.

Nishimune, H., Sanes, J.R., and Carlson, S.S. (2004). A synaptic laminin-calcium channel interaction organizes active zones in motor nerve terminals. *Nature,* 432, 580-587.

Nishimune, H., Valdez, G., Jarad, G., Moulson, C.L., Müller, U., Miner, J.H., and Sanes, J.R. (2008). Laminins promote postsynaptic maturation by an autocrine mechanism at the neuromuscular junction. *J. Cell Biol.,* 182, 1201-1215.

Noakes, P.G., Gautam, M., Mudd, J., Sanes, J.R., and Merlie J. P. (1995). Aberrant differentiation of neuromuscular junctions in mice lacking s-laminin/laminin beta 2. *Nature*, 374, 258-262.

Ochi, H., and Westerfield, M. (2007). Signaling networks that regulate muscle development: lessons from zebrafish. *Dev. Growth Differ.,* 49, 1-11.

Parsons, M.J., Pollard, S.M., Saude, L., Feldman, B., Coutinho, P., Hirst, E.M. and Stemple, D. L. (2002). Zebrafish mutants identify an essential role for laminins in notochord formation. *Development,* 129, 3137-3146.

Patton, B.L., Chiu, A.Y., and Sanes, J.R. (1998). Synaptic laminin prevents glial entry into the synaptic cleft. *Nature,* 393, 698-701.

Patton, B.L., Cunningham, J.M., Thyboll, J., Kortesmaa, J., Westerblad, H., Edstrom, L., Tryggvason, K., and Sanes J. R. (2001). Properly formed but improperly localized synaptic specializations in the absence of laminin alpha4. *Nat. Neurosci.,* 4, 597–604.

Paulsson, M., Saladin, K., and Engvall, E. (1991). Structure of laminin variants. The 300-kDa chains of murine and bovine heart laminin are related to the human placenta merosin heavy chain and replace the A chain in some laminin variants. *J. Biol. Chem.,* 266, 17545–17551.

Petäjäniemi, N., Korhonen, M., Kortesmaa, J., Tryggvason, K., Sekiguchi, K., Fujiwara, H., Sorokin, L., Thornell, L.E., Wondimu, Z., Assefa, D., Patarroyo, M., and Virtanen, I. (2002). Localization of laminin α4-chain in developing and adult human tissues. *J. Histochem. Cytochem.,* 50, 1113-1130.

Pourquié, O. (2001). Vertebrate somitogenesis. *Annu. Rev. Cell Dev. Bio.,* 1 17, 311-50.

Qiao, C., Li, J., Zhu, T., Draviam, R., Watkins, S., Ye, X., Chen, C., and Xiao, X. (2005). Amelioration of laminin-alpha2-deficient congenital muscular dystrophy by somatic gene transfer of miniagrin. *Proc. Natl. Acad. Sci. USA,* 102, 11999–12004.

Relaix, F., and Zammit, P.S. (2012). Development Satellite cells are essential for skeletal muscle regeneration: the cell on the edge returns centre stage. *Development,* 139, 2845-2856.

Rooney, J.E., Gurpur, P.B., Yablonka-Reuveni, Z., and Burkin, D.J. (2009a). Laminin-111 restores regenerative capacity in a Congenital Myopathy. *Am. J. Pathol.,* 174, 256–264.

Rooney, J.E., Gurpur, P.B., and Burkin, D.J. (2009b). Laminin-111 protein therapy prevents muscle disease in the mdx mouse model for Duchenne muscular dystrophy. *Proc. Natl. Acad. Sci. USA,* 106, 7991-6.

Rooney, J.E., Knapp, J.R., Hodges, B.L., Wuebbles, R.D., and Burkin, D.J. (2012). Laminin-111 Protein Therapy Reduces Muscle Pathology and Improves Viability of a Mouse Model of Merosin-Deficient Congenital Muscular Dystrophy. *Am. J. Pathol.,* 180,1593-1602.

Rozario, T., and DeSimone, D. W. (2010). The extracellular matrix in development and morphogenesis: A dynamic view. *Dev. Biol.,* 341, 126-140.

Rybakova, I.N., Patel, J.R., and Ervasti, J.M. (2000). The dystrophin complex forms a mechanically strong link between thesarcolemma and costameric actin. *J. Cell Biol.,* 150, 1209-1214.

Sanes, J.R. (2003). The basement membrane/basal lamina of skeletal muscle. *J. Biol. Chem.,* 278, 12601-12604.

Schuler, F., and Sorokin, L.M. (1995). Expression of laminin isoforms in mouse myogenic cells in vitro and in vivo. *J. Cell Sci.,* 108, 3795-805.

Schultz, E., and McCormick, K.M. (1994) Skeletal muscle satellite cells. *Rev. Physiol. Biochem. Pharmacol.,* 123, 213–257.

Shibuya, S., Wakayama, Y., Inoue, M., Kojima, H., and Oniki, H. (2003). Merosin (laminin-2) localization in basal lamina of normal skeletal muscle fibres and changes in plasma membrane of merosin-deficient skeletal muscle fibres. *Med. Electron. Microsc.,* 36, 213-20.

Silva-Barbosa, S.D., Butler-Browne, G.S., de Mello, W., Riederer, I., Di Santo, J.P., and Savino, W. (2008). Human myoblast engraftment is improved in laminin-enriched microenvironment. *Transplantation,* 85, 566–575.

Smirnov, S.P., Barzaghi, P., McKee, K.K., Ruegg, M.A., and Yurchenco, P.D. (2005). Conjugation of LG domains of agrins and perlecan to polymerizing laminin-2 promotes acetylcholine receptor. *clustering J. Biol. Chem.,* 280, 41449-41457.

Sorokin, L.M., Pausch, F., Durbeej, M., and Ekblom, P. (1997). Differential expression of five laminin α (1–5) chains in developing and adult mouse kidney. *Dev. Dyn.,* 210, 446–462.

Sunada, Y., Bernier, S.M., Kozak, C.A. Yamada, Y., and Campbell, K.P. (1994). Deficiency of merosin in dystrophic dy mice and genetic linkage of laminin M chain gene to dy locus. *J. Biol. Chem.,* 269, 13729–13732.

Sztal, T., Berger, S., Currie, P.D., and Hall, T.E. (2011). Characterization of the laminin gene family and evolution in zebrafish. *Dev. Dyn.,* 240, 422-431.

Talts, J.F., Sasaki, T., Miosge, N., Göhring, W., Mann, K., Mayne R, and Timpl, R. (2000) Structural and functional analysis of the recombinant G domain of the laminin alpha4 chain and its proteolytic processing in tissues. *J. Biol. Chem.,* 275, 35192-35199.

Timpl, R., Tisi, D., Talts, J.F., Andac, Z., Sasaki, T., and Hohenester, E. (2000). Structure and function of laminin LG modules. *Matrix Biol.,* 19, 309-317.

Tomé, F.M., Evangelista, T., Leclerc, A., Sunada, Y., Manole, E., and Estournet, B. (1994). Congenital muscular dystrophy with merosin deficiency. *CR Acad. Sci.,* II 317, 351-357.

Urbano, J.M., Torgler, C.N., Molnar, C., Tepass, U., López-Varea, A., Brown, N.H., de Celis, J.F., and Martín-Bermudo, M.D. (2009). Drosophila laminins act as key regulators of basement membrane assembly and morphogenesis. *Development,* 136, 4165-4176.

Vachon, P.H., Loechel, F., Xu, H., Wewer, U.M., and Engvall, E. (1996). Merosin and laminin in myogenesis; specific requirement for merosin in myotube stability and survival. *J. Cell Biol.,* 134, 1483-1497.

Vainzof, M., Ayub-Guerrieri, D., Onofre, P.C., Martins, P.C., Lopes, V.F., Zilberztajn, D., Maia, L.S., Sell, K., and Yamamoto, L.U. (2008). Animal models for genetic neuromuscular diseases. *J. Mol. Neurosci.,* 34, 241–248.

Vivarelli, E., Brown, W. E., Whalen, R. G. and Cossu, G. (1988). The expression of slow myosin during mammalian somitogenesis and limb bud differentiation. *J. Cell Biol.,* 107, 2191-2197.

Von der Mark, K., and Ocalan, M. (1989). Antagonistic effects of laminin and fibronectin on the expression of the myogenic phenotype. *Differentiation,* 40, 150-157.

Wedlich, D., Hacke, H., and Klein, G. (1989). The distribution of fibronectin and laminin in the somitogenesis of Xenopus laevis. *Differentiation,* 40, 77-83.

White, J.G., Southgate, E., Thomson, J.N., and Brenner, S. (1976). The structure of the ventral nerve cord of Caenorhabditis elegans . *Philos. Trans . R. Soc. London Ser. B. Biol.,* 275, 327-348.

Wu, H., Xiong, W.C., and Mei, L. (2010). To build a synapse: signaling pathways in neuromuscular junction assembly. *Development,* 137, 1017-1033.

Xu, H., Wu, X.R., Wewer, U.M., and Engvall, E. (1994a). Murine muscular dystrophy caused by a mutation in the laminin $\alpha 2$ (Lama2) *gene. Nat. Genet.,* 8, 297–302.

Xu, H., Christmas, P., Wu, X. R., Wewer, U. M., and Engvall, E. (1994b). Defective muscle basement membrane and lack of M-laminin in the dystrophic dy/dy mouse. *Proc. Natl. Acad. Sci. USA,* 91, 5572–5576.

Yarnitzky, T., and Volk, T. (1995). Laminin is required for heart, somatic muscles, and gut development in the Drosophila embryo. *Dev. Biol.,* 169, 609-618.

Yurchenco, P.D., and Patton, B.L. (2009). Developmental and pathogenic mechanisms of basement membrane assembly. *Curr. Pharm. Des.,* 15, 1277-1294.

Yurchenco, P.D.. (2011). Basement membranes: cell scaffoldings and signaling platforms. *Cold Spring Harb. Perspect. Biol.,* 3, a004911.

Zammit, P.S., Partridge, P.A., and Yablonka-Reuveni, Z. (2006). The skeletal muscle satellite cells: the stem cell that came in from the cold. *J. histochem. Cytochem,* 54, 1177-1191.

In: Laminins: Structure, Biological Activity and Role in Disease ISBN: 978-1-62618-568-5
Editors: Demetrius C. Adams and Emil O. Garcia

Chapter 2

Laminin-332 and Integrins: Signaling Platform for Cell Adhesion and Migration and Its Regulation by *N*-glycosylation

***Yoshinobu Kariya*[*1], *Yukiko Kariya*[2] *and Jianguo Gu*[3]**

[1]Department of Biochemistry,
[2]Department of Diagnostic Pathology,
Fukushima Medical University School of Medicine,
Hikarigaoka, Japan
[3]Division of Regulatory Glycobiology,
Institute of Molecular Biomembrane and Glycobiology,
Tohoku Pharmaceutical University,
Aoba-ku, Sendai, Japan

Abstract

Laminin-332 (formerly known as laminin-5) is a major component of basement membranes in the skin and other stratified squamous epithelial tissues. Laminin-332 has attracted much attention from both physicians and researchers as a disease-related and functional molecule. The study of human genetic disorders and a knockout mouse model has provided the evidence that laminin-332 is indispensable for adhesive structures in the skin. In wound healing and cancer invasion, overexpression of laminin-332 accelerates cell migration by activating PI3K, PKC, and MAPK signaling pathways. These functional activities of laminin-332 are through interactions with several cell surface receptors including integrins $\alpha3\beta1$ and $\alpha6\beta4$, and syndecans. In addition, recent studies revealed that posttranslational modifications, such as proteolytic processing and *N*-glycosylation, of laminin-332 and of integrins $\alpha3\beta1$ and $\alpha6\beta4$ are important for the

* Corresponding author: Department of Biochemistry, Fukushima Medical University School of Medicine, 1 Hikarigaoka, Fukushima 960-1295, Japan. Tel: +81-24-547-1144; Fax: +81-24-548-8641; E-mail: kariya@fmu.ac.jp.

laminin-332 activities. This review summarizes the molecular mechanisms and significance of cellular adhesion and migration through laminin-332 as well as integrins $\alpha3\beta1$ and $\alpha6\beta4$. Understanding of these regulatory mechanisms could lead to development of new molecular targeted therapies, particularly in oncology.

Introduction

Basement membrane (BM) is a 50–100-nm layer of specialized extracellular matrix (ECM) protein complex that underlies all epithelial cell sheets and tubes, such as epithelia, endothelia, muscle cells, fat cells, and nerve cells [1]. The major ECM components of BM are laminin, type IV collagen, nidogen, and perlecan. Self-assembly of laminin and type IV collagen individually forms insoluble networks, which are linked by nidogen and perlecan, and the resulting supramolecular structures provide BM stability and mechanical support of tissue. BMs not only maintain tissue structure but also regulate cellular functions such as cell adhesion, migration, differentiation, proliferation, and resistance to anoikis [2]. These BM-mediated cellular functions are dependent on the type of ECM components in BM.

Laminins are large extracellular glycoproteins with a molecular mass of 400–900 kDa [2]. Laminins are heterotrimers consisting of three chains, α, β, and γ, bound by disulfide bonds (Figure 1). To date, five α, four β, and three γ chains as well as alternatively spliced variants of these chains have been genetically identified to form at least 16 distinct laminin heterotrimers in mammals, although no laminin heterotrimer containing the $\beta4$ chain has been described [3]. These laminin isoforms show tissue- and development-specific expression patterns *in vivo* and lead to distinct and tissue-specific functions of BM. For example, laminin-211 ($\alpha2\beta1\gamma1$; formerly laminin-2) and laminin-411 ($\alpha4\beta1\gamma1$; formerly laminin-4) is predominantly expressed in BMs of developing skeletal muscles and peripheral nerves and in BMs of blood vessels, respectively. The first identified laminin, laminin-111 ($\alpha1\beta1\gamma1$; formerly laminin-1), which was discovered in mouse Engelbreth-Holm-Swarm (EHS) sarcoma, is primarily expressed during very early embryogenesis [2-4].

The laminin $\alpha3$ chain is predominantly expressed in the BM of skin and other stratified squamous epithelial tissues and is known to have two splicing variants: the truncated form α3A and the full-sized form α3B [5, 6]. Laminin α3A and α3B chains associate with the $\beta3$ chain and the $\gamma2$ chain to form laminin-332 (laminin-3A32; formerly laminin-5 or laminin-5A) and laminin-3B32 (formerly laminin-5B), respectively [7]. The laminin α3A chain can also associate with the $\beta1/\beta2$ chains and the $\gamma1$ chain, forming laminin-311 (laminin-3A11; formerly laminin-6 or laminin-6A) and laminin-321 (laminin-3B21; formerly laminin-7 or laminin-7A) [8, 9]. The laminin α3B chain is widely expressed in vascular BMs of normal tissues, probably as laminin-3B11/3B21 (laminin-6B/7B) but the expression of laminin α3B chain is lost in vessels near invasive breast carcinoma cells [10-12].

Compared with laminin-111, all short arms in laminin-332 subunits are truncated (Figure 1). This truncated structure is unique to laminin-332. Although full sized laminin short arms contain some integrins and heparin-binding sites as well as self-polymerization sites, truncated forms of laminin-332 would lose a part of these domains, and consequently would have distinct physiological properties from other laminin isoforms. In contrast, all laminin α chains have a central long coiled-coil domain and a large C-terminal comprising five tandem laminin globular (LG) subdomains (LG1-5) [2]. The latter is the major interaction sites for cell surface receptors such as integrins, syndecans, and dystroglycans (Figure 1).

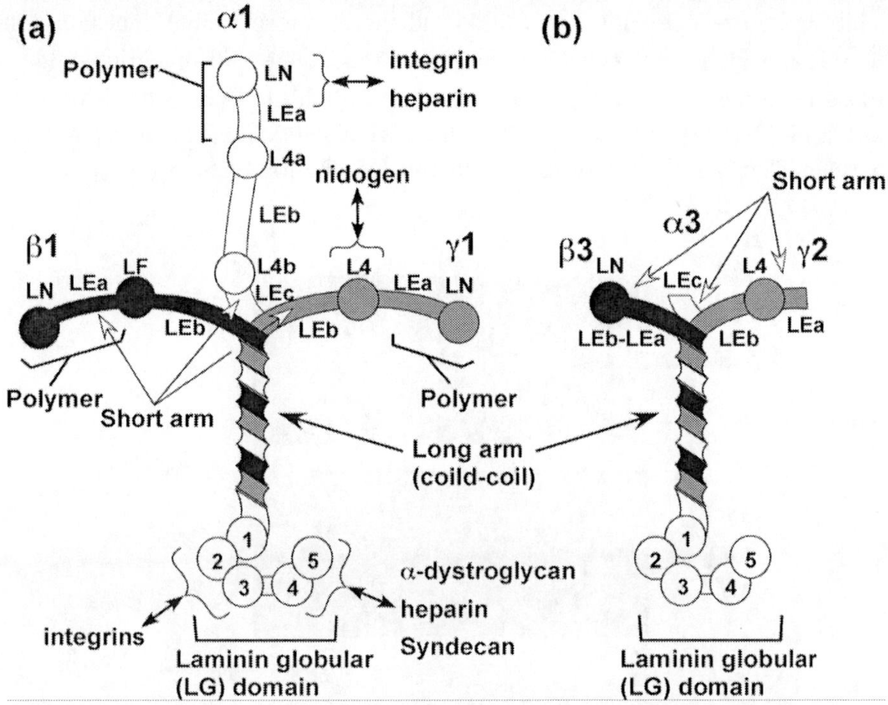

Figure 1. Schematic structures of laminin-111 and laminin-332. Laminins are heterotrimers composed of α, β, and γ chains bound by disulfide bonds. Each laminin chain contains a short arm and a long arm, forming coiled-coil domains. Only the α chain has five tandem laminin globular (LG) subdomains (LG1-5) at its carboxyl-terminal end. The amino-terminal (LN domain) and internal short arm globular (LF and L4 domains) modules are indicated by ovals. The rod-like epidermal growth factor (EGF) repeats (LE domains) are shown as lines. (a) Laminin-111 (α1β1γ1) can self-polymerize through the three LN domains, possibly requiring the participation of more distal domains in the short arms. Heparin- and integrin-binding sites are found in both the N-terminal LN domain and the C-terminal LG domain of the α1 chain. (b) Compared with laminin-111, all short arms in laminin-332 (α3β3γ2) subunits are truncated. Since the N-terminal LN domain in the laminin α1 chain contains some integrin- and heparin-binding sites as well as self-polymerization sites, laminin-332 would consequently show distinct functional activities from laminin-111. The protein binding partners of laminin-332 are described in Figures 2B and 3.

Laminin-332 in Cell Adhesion and Deposition

As shown in human genetic disorders and a knockout mouse model, laminin-332 is an essential molecule for maintaining skin tissue architecture. Laminin α3, β3, and γ2 chain knockout mice die at the neonatal stage and suffer blistering of the skin [13-15]. These phenotypes are consistent with the human severe and lethal skin blistering disease, Herlitz's junctional epidermolysis bullosa, which is caused by genetic defects of any of the laminin-332 subunits [16-19]. Laminin-332 is predominantly expressed in the BM of skin as well as other stratified squamous epithelial tissues (Figure 2A). Skin tissue is separated into two parts epidermis and dermis by laminin-332–containing BM, and only basal keratinocytes can associate with BM (Figure 2B). Basal keratinocytes associate with laminin-332 mainly through cell surface receptor integrin α6β4, and thereby form stable adhesion complexes, termed hemidesmosomes [20, 21]. Laminin-332 can associate with integrin α6β4 and other hemidesmosome-related protein, type VII collagen through the LG1–3 domains [22] and β3 chain short arm [23, 24], respectively (Figure 2B). Genetic defects in such laminin-332 binding partners also cause other forms of epidermolysis bullosa [25-28].

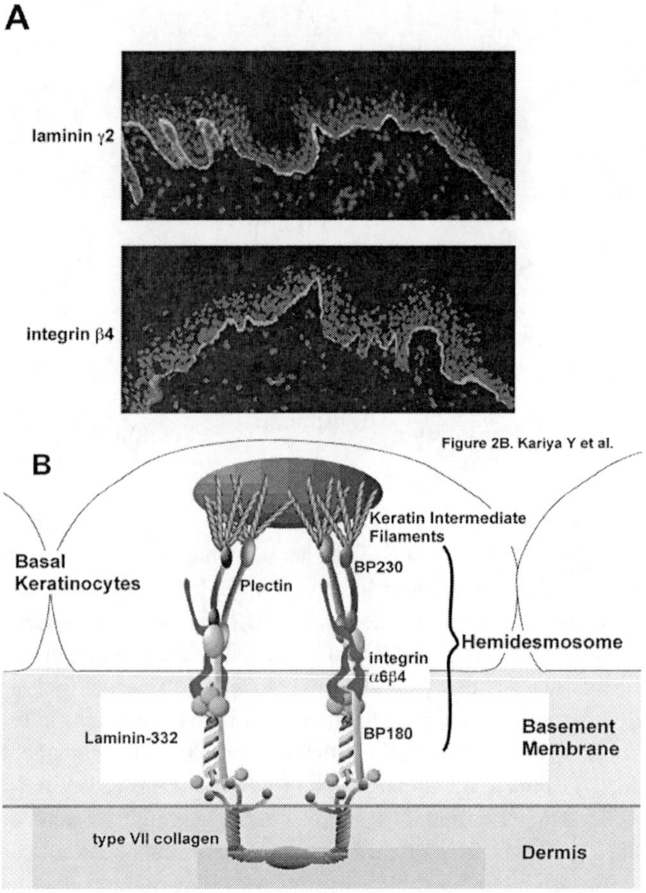

Figure 2. Hemidesmosome components in the skin. (A) Immunohistochemical staining of the laminin γ2 subunit and integrin β4 subunit in the skin. Both laminin γ2 (upper panel) and integrin β4 (lower panel) are localized in the basement membrane. Nuclei are stained with Hoechst 333432. (B) A schematic diagram of hemidesmosomes in the skin. Laminin-332 binds to integrin α6β4 and type VII collagen through the laminin globular (LG) domains 1–3 in the α3 chain and the short arm in the β3 chain, respectively. BP180 associates with laminin-332 and α6 integrin through the extracellular domain and with integrin β4, plectin, and BP230 through the cytoplasmic domain. Plectin and BP230 bind to the β4 cytoplasmic domain as well as keratin intermediate filaments. Genetic defects involving these molecular components of the hemidesmosomes cause epidermolysis bullosa.

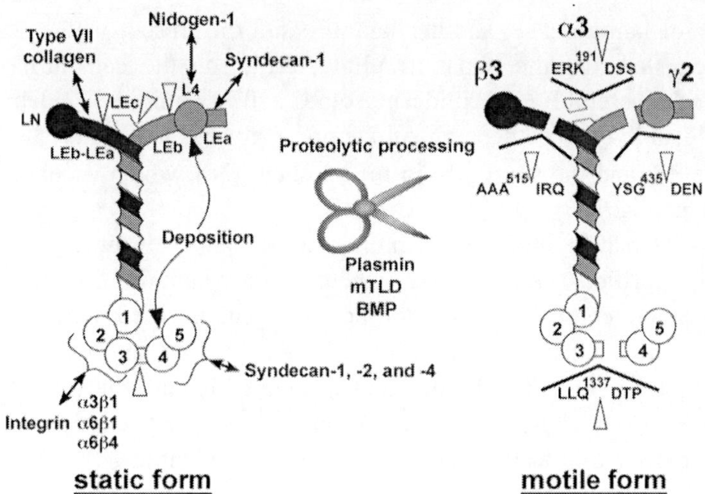

Figure 3. Regulatory mechanism of laminin-332 activities by proteolytic processing. The LG1–3 domains in the α3 chain bind to integrins α3β1, α6β1, and α6β4, and the LG4–5 domains interact with syndecans. The LEb–LEa domain in the β3 chain binds to type VII collagen. The L4 domain and LEa domain in the γ2 chain binds to nidogen-1 and syndecan-1, respectively. The precursor α3 chain (190 kDa) is cleaved in the linker region between the LG3 and LG4 domains by plasmin, mammalian tolloid (mTLD), and bone morphogenic protein-1 (BMP-1; 160 kDa). In addition to the proteolytic processing at the C-terminal G domain, in some cell lines, further proteolytic processing occurs at an N-terminal LEc domain in the α3 chain (145 kDa). The γ2 chain undergoes processing in the LEb domain by mTLD and BMP-1. Although the β3 chain is relatively resistant to proteolysis compared with the α3 and γ2 chains, the β3 chain is also cleaved at the short arm region. Arrowheads indicate the major proteolytic cleavage sites. The numbers and amino acid sequences indicate the position of amino acid residues at the cleavage sites and the amino acid sequence around the cleavage sites. The α3 LG4–5 domains and the γ2 L4 domain play critical roles in the incorporation of laminin-332 into the extracellular matrix, thereby supporting cell adhesion and suppressing cell migration (static form). Proteolytic processing of laminin-332 suppresses its deposition and its association with other molecules, thereby inducing cell migration (motile form).

Recent analysis using electron microscopy showed that the deposited laminin-332 matrix forms a mesh-like network structure and is highly polymerized [29]. Full-sized laminin chains, such as laminin α1, α5, β1, and γ1 chains, can self-assemble through the self-polymerization site in the short arm LN domains [2] (Figure 1). In contrast, because the LN domain is truncated in laminin α3 and γ2 chains, laminin-332 has been considered to be incapable of self-polymerizing and assembling into BM networks. Although the mechanism has not yet been clarified, approximately half of total laminin-332 exists in disulfide-linked complexes with laminin-311 or laminin-321 formed through the unpaired cysteine residues in their short arms [9]. Because laminins-311/321 have full-sized β1 or γ1 chains containing the self-polymerization site, laminin-332 could be integrated into a matrix network through laminins-311/321.

During wound healing, newly synthesized laminin-332 is deposited under cells as an insoluble matrix protein [30]. Deposition of laminin-332 to the matrix is facilitated by the LG4–5 domains in α3 chain [31, 32] and the L4 domain in γ2 chain [33] (Figure 3). In contrast, the short arm of the laminin β3 chain enhances the matrix assembly and cell adhesion activity of laminin-511, another laminin isoform present in the skin [34]. Recent studies have suggested that the heparan sulfate chains on the cell surface also play an important role in the assembly of hemidesmosomes and in laminin-332 deposition [29, 35]. This may be reasonable because heparan sulfate proteoglycans bind to the LG4–5 domains in the α3 chain [36, 37] and the short arm in the γ2 chain [38], which accelerate deposition of laminin-332 into the matrix.

The laminin-332 matrix supported keratinocyte cell adhesion much more strongly than non-polymerized, purified laminin-332, whereas the laminin-332 matrix suppressed keratinocyte migration compared with non-polymerized, purified laminin-332 [29]. This strong adhesion of the laminin-332 matrix to cells is caused by the high affinity of laminin-332 to integrin α3β1, and an induction of α6β4-containing hemidesmosome-like structures in the laminin-332 matrix. These observations suggest that polymerized lamnin-332 has different physiological activities from unassembled soluble laminin-332.

Integrins α3β1 and α6β4 in Cell Adhesion

Integrins are a large family of heterodimeric transmembrane receptors comprising α and β subunits [39]. In mammals, 18 α and 8 β subunits have been characterized, and the combinations of them form 24 distinct integrins. The extracellular domains of integrin subunits associate with ECM proteins, such as laminin, collagen, fibronectin, and vitronectin, and in the cytoplasmic domain many cellular signaling molecules bind to integrins and send various cellular signaling. Thus, integrins constitute both a structural connection and a bi-directional signaling pathway that crosses the cell membrane [39, 40]. The laminin-binding integrins (α3β1, α6β1, α6β4, and α7β1) show robust association with tetraspanin proteins [41]. Tetraspanins CD151, CD81, and CD9 modulate laminin binding, affecting integrin-dependent neurite outgrowth, cell adhesion, migration, and morphology [41-43].

The main integrin receptors for laminin-332 are integrins α3β1, α6β1, and α6β4, which are highly expressed on epithelial cells as well as on cancer cells [22, 44, 45]. *In vitro* cell adhesion activity of laminin-332 through these integrins is much stronger than that of other ECM proteins such as other laminin isoforms, collagens, fibronectin, and vitronectin [46, 47]. The Glu residue at the third position from the C-terminus of the γ2 chain is critical for the association of integrins α3β1 and α6β1 with laminin-332 [48]. Keratinocytes express integrins α2β1, α3β1, and α6β4 but not α6β1 [20], and integrin α6β4 regulates α2 and α3 integrin subunit expression [49]. Integrins α3β1 and α6β4 function as cell adhesion receptors for laminin-332. The association between those integrins and laminin-332 is important for maintaining the integrity of epidermal structure and function.

Integrin α3β1-mediated adhesion to laminin-332 promotes the stabilization of leading lamellipodia in migrating keratinocytes through activation of Rac1 [50]. In contrast to the localization of integrin α6β4 in hemidesmosome, integrin α3β1 is recruited to focal contacts in cultured cells [21]. Because integrin α3β1 but not α6β4-mediated cell spreading and adhesion to laminin-332 are dependent on the cAMP-Epac-Rap1 pathway [51], the regulatory mechanisms for cell spreading and adhesion between integrins α3β1 and α6β4 are distinct. Patients with homozygous mutations in *ITGA3* gene suffer from congenital nephrotic syndrome, interstitial lung disease, and epidermolysis bullosa [52]. Similar to the phenotypes of human patients, in integrin α3β1-deficient mice, matrix disorganization is accompanied by skin blistering [53]. However, epidermal blistering was caused by rupture of BM rather than by detachment of the epidermis from BM. Integrin α3β1-deficient keratinocytes from α3β1-deficient mice attach to laminin-332 through integrin α6β4 but spread poorly on laminin-332. Integrin α3β1 therefore is considered to function in cell spreading and migration rather than stable adhesion.

Integrin α3β1 is also found in cell–cell contacts in adherence junctions [54] and regulates cadherin–catenin complex formation [55]. Association of integrin α3β1 with tetraspanin CD151 stimulates cadherin-mediated cell–cell adhesion both by regulating the expression of PTPμ and by organizing the multimolecular association of PKCβII, RACK1, PTPμ, E-cadherin, and β-catenin [56]. Integrin α3β1 also regulates cell–cell adhesion by promoting gap junctional intercellular communication through interaction with laminin-332 [57]. Furthermore, integrin α3β1 plays important roles in proper incorporation of laminin-332 into the matrix and in the expression of the laminin-332 matrix function [53, 58]. This integrin α3β1-mediated laminin-332 deposition is regulated by the Rac activator Tiam1 [59].

Integrin α6β4 is an essential component of hemidesmosomes [20, 60] (Figure 2A and 2B). Patients with mutations in the genes encoding integrin α6β4 (*ITGA6* and *ITGB4*) suffer from epidermolysis bullosa with pyloric atresia (EB-PA), which is an autosomal recessive disorder characterized by blistering of the skin and mucous membranes and by associated congenital gastrointestinal atresia [25, 28]. Unlike other β integrins, the cytoplasmic tail of β4 integrin is more than 1000 amino acids long and contains two pairs of fibronectin type III (FNIII) repeats separated by a connecting segment [60]. This long cytoplasmic region of β4 integrin associates with hemidesmosome component molecules, plectin, BP180 (type XVII collagen or BPAG2), and BP230 (BPAG1) [60] (Figure 2B). Two missense mutations (R1225H and R1281W) in the β4 gene (*ITGB4*) lead to non-lethal phenotypes of EB-PA [25, 61]. R1225 and R1281 in the first pair of FNIII repeats are essential for binding of integrin β4 to plectin. In R1225H or R1281W β4 mutants expressing β4 null keratinocytes obtained from EB-PA patients, mutations in β4 abolish the interaction between β4 integrin and plectin and prevent the recruitment of plectin into hemidesmosomes [62]. The binding of β4 integrin to plectin is a critical step in the formation of hemidesmosomes as well as in the recruitment of BP180 and BP230 into hemidesmosomes [63].

Laminin-332 and Integrins in Cell Motility

In vitro, laminin-332 stimulates the migration of various types of cells, including cancer cells, and is known not only as a cell substrate but also as a soluble cell scattering factor [64-66]. *In vivo*, laminin-332 is overexpressed during wound healing [6] and cancer invasion [67-69] and is considered to be a key factor for such physiological events. *In vitro* cell migration, wound healing, and invasion assays using an antibody that blocks the function of laminin-332 or laminin-332–deficient keratinocytes indicated that laminin-332 is an essential molecule in cell motility [70, 71]. Furthermore, both the binding of α6β4 integrin to laminin-332 and the cytoplasmic domain of β4 integrin are required for the EGF-driven directional migration of keratinocytes [72]. Laminin-332 promotes both cell adhesion and migration through the interaction of the LG3 domain in the C-terminus of the laminin α3 chain with integrins α3β1, α6β1, and α6β4 [22]. The apparent discrepancy between strong cell adhesion and migration activities can be explained by the presence of distinct functional sites for these two activities within the LG3 domain [22]. The association of the LG3 domain with integrins activates cellular signaling through protein kinase C, phosphatidylinositol 3-kinase (PI3K), Rac, ERK, JNK, and NF-kB, thereby promoting cell migration [65, 73-75].

Human laminin-332 is synthesized and secreted as a precursor form (190 kDa α3, 135 kDa β3 and 150 kDa γ2 chains). In both physiological and pathological conditions, such as wound healing and cancer invasion, laminin-332 undergoes proteolytic processing by some proteases after secretion by cells (Figure 3). In fact, after an injury, leading keratinocytes migrate into the wound bed where dermal collagens are exposed and secrete large amounts of laminin-332. Laminin-332 containing the precursor form of the α3 chain was found only in wound beds, whereas the mature form of laminin-332 was detected both in normal tissue, away from the wound bed, and in the provisional BM of the wound bed [30]. Proteolytic processing converts the precursor form of α3 (190 kDa) and γ2 (150 kDa) chains to the mature form of α3 (160 kDa and 145 kDa) and γ2 (105 kDa) chains, respectively [76] (Figure 3). The α3 chain of laminin-332 is cleaved in the linker region between the LG3 and LG4 domains (160 kDa) [77] by plasmin [78], mammalian tolloid (mTLD), and bone morphogenic protein-1 (BMP-1) [79, 80]. In addition to proteolytic processing at the C-terminal G domain, in some cell lines, further proteolytic processing occurs at an N-terminal LEc domain in the α3 chain (145 kDa) [7]. In human laminin-332, mTLD and BMP-1 are major processing enzymes of the γ2 chain [79, 80], which cleave the γ2 chain at the short arm LEb domain, although in rat laminin-332, matrix metalloproteinase 2 (MMP2) [81], membrane type 1-matrix metalloproteinase (MT1-MMP) [82], and neutrophil elastase [83] are also known to process the γ2 chain. Although the β3 chain is relatively resistant to proteolysis compared with the α3 and γ2 chains, the β3 chain is also cleaved at the short arm region [24, 84] by MT1-MMP [84], matrilysin-1 (matrix metalloproteinase 7; MMP7) [85], and hepsin [86].

These proteolytic conversions of laminin-332 affect its activities. Laminin-332 containing the mature form of the γ2 chain (105 kDa) increases cell migration but decreases cell adhesion, compared with laminin-332 containing the precursor form of γ2 chain (150 kDa) [81, 82, 87]. Purified recombinant laminin-332 containing the precursor form of α3 chain decreases both cell adhesion and migration compared with recombinant laminin-332 containing the mature form of the α3 chain that lacks the LG4–5 domains [32]. The main reason for the downregulation of biological activities in laminin-332 with LG4–5 domains may be decreased integrin-mediated signaling through the LG1–3 domains because of interference by the association of the LG4–5 domains with heparan sulfate proteoglycan receptors such as syndecans-1, 2, and 4 [88].

The integration of laminin-332 into ECM supports stable cell adhesion and suppresses cell migration by strong association with the integrins α3β1 and α6β4 and other matrix protein [29]. Dissociation from stable adhesion constituted by deposited laminin-332 therefore is required for inducing cell migration. The laminin-332 with LG4–5 domains is integrated into ECM more efficiently than its mature form because of the interaction of LG4–5 domains with heparan sulfate proteoglycans such as syndecans [31, 32, 37] (Figure 3).

Similarly, the N-terminal L4 domain of the γ2 chain is participated in the integration of laminin-332 into ECM [33] (Figure 3). Proteolytic processing of α3 and γ2 chains remove the LG4–5 domains from the α3 chain and the L4 domain from the γ2 chain in the laminin-332 molecule, resulting in suppression of laminin-332 deposition and its association with other molecules, thereby inducing cell migration. The LEb–LEa domain in the short arm region of the β3 chain associates with type VII collagen [23, 24], and the L4 domain and LEa domain of the γ2 chain interacts with nidogen-1 [89] and syndecan-1 [38], respectively. Because type VII collagen is the main component of anchoring fibrils and nidogen is a cross-linking protein among other matrix molecules, reduction of the association of laminin-332 with these molecules by proteolytic processing at the short arm region of the β3 [24, 84-86] and γ2 [81, 82] chains increases laminin-332–dependent cell motility.

During cell migration, the hemidesmosome assembly/disassembly is dynamically regulated [60, 90]. Serine/threonine phosphorylation of the cytoplasmic domain of β4 integrin by stimulation of growth factors, such as EGF, destabilizes hemidesmosomes [91]. Hemidesmosome formation is driven by interaction of the cytoplasmic domain of β4 integrin with plectin. The interaction between plectin and β4 integrin is regulated by serine phosphorylation of residues 1356, 1360, 1364, and 1424 [92] or threonine phosphorylation of residue 1736 [93] of the β4 cytoplasmic domain. The phosphorylation of these residues by PKC, PKA, or PKD1, which signal downstream of epidermal growth factor receptor (EGFR) results in hemidesmosome disassembly due to loss of interaction between β4 integrin and plectin, thereby inducing cell migration.

Laminin-332 and Integrins in Cancer

Many studies have reported that laminin-332 [66, 67] and its integrin receptors [94] are highly expressed in tumors, and previous *in vitro* studies of laminin-332 activities, such as cell adhesion, migration, and proliferation, indicated that laminin-332 and its integrin receptors are key factors in tumor progression [95]. The importance of laminin-332 and integrin α6β4 in tumorigenesis has been proven using an animal model of human squamous cell carcinoma (SCC). In the assay, transformed primary keratinocytes could form tumors in nude mice skin, but laminin-332–negative or β4 integrin-negative keratinocytes (derived from epidermolysis bullosa patients with *LAMB3* or *ITGB4* null mutations, respectively) could not [96]. Introduction of laminin β3 and integrin β4 cDNA expression vectors into those cells restored their tumorigenicity, suggesting that these molecules are essential for human SCC tumorigenesis. More extensive studies about laminin-332–mediated tumorigenesis have shown that the laminin α3 LG4–5 domains and the laminin β3 LEb–LEa domain are significant for SCC tumorigenesis. The LG4–5 domains activate ERK and PI3K signaling pathways that induce proliferation and inhibit apoptosis, promoting SCC tumorigenesis [97]. The association of the laminin β3 LEb–LEa domain with the type VII collagen NC1 domain, which promotes tumor cell invasion in a laminin-332–dependent manner, stimulates PI3K signaling for SCC tumorigenesis [98].

Integrin α6β4 was originally identified as a tumor-specific protein [99, 100]. Subsequent studies have reported that overexpression of integrin α6β4 was seen in several types of invasive and metastatic carcinomas and correlated with poor prognosis [101]. Integrin α6β4 activates PKC [102], PI3K [103], and ERK [104] signaling pathways, which are linked to cell migration, invasion, and proliferation [95]. In particular, PI3K, which is an upstream regulator of Rac1, is the central factor in α6β4-dependent tumor progression [103, 105]. Integrin α6β4 associates with several receptor tyrosine kinases (RTKs), including EGFR, ErbB2, Met, and Ron [95]. These RTKs activate β4 integrin-associated Src family kinases and thereby induce the tyrosine phosphorylation of the β4 integrin cytoplasmic domain. The phosphorylation of tyrosine 1494 in the β4 cytoplasmic domain promotes breast carcinoma cell invasion through the combined activation of PI3K and Src, and anchorage-independent growth through activation of ERK1/2 [106]. The phosphorylated α6β4 integrin conversely promotes Src-family kinases-dependent phosphorylation of RTKs. ErbB2, the ligand-less member of EGFR family, associates with integrin β4 subunit through a region of the β4 cytoplasmic domain and both α6β4 and ErbB2 are required for PI3K activation and subsequent invasion in mouse NIH3T3 cells [107]. Conversely, the binding of immunoglobulin-like adhesion molecule Necl-2 (Nectin-like molecule 2) to the extracellular region of β4 integrin stabilizes the hemidesmosome structure, thereby inhibiting cancer cell motility by suppressing ErbB3/ErbB2 signaling [108].

Integrin α3β1 is also associated with tumor progression in some cancers. In hepatocellular carcinoma (HCC) cells but not in peritumoral tissue of the same HCC patients, laminin-332, Snail, and Slug are upregulated, whereas E-cadherin is downregulated and β-catenin is translocated into the nuclei. *In vitro* incubation of HCC invasive cells with purified laminin-332 induces upregulation of Snail and Slug and downregulation of E-cadherin as well as translocation of β-catenin into nuclei, resulting in a complete EMT accompanied by marked morphological change. This event is mediated through integrin α3β1 but not α6β1 or α6β4 [109]. Another group, using integrin α3-null kidney epithelial cells, demonstrated that integrin α3β1 attenuates EMT by downregulating β-catenin–Smad2 signaling when α3β1 is mainly engaged in laminin-332–rich rat 804G culture supernatant or with E-cadherin in adherens junctions [110]. These apparent contradictory findings may result from the laminin-332 source or the cell types used in each assay.

Regulation of Laminin-332 and Integrins by *N*-Glycosylation

Laminin is a heavily glycosylated molecule and between 13% and 30% of its total molecular weight is contributed by *N*-glycosylation [111]. The laminins undergo terminal glycosylation within the Golgi apparatus and then *N*-glycosylated laminins are secreted by the cells. In early studies, the carbohydrate structure and function of laminin-111 was intensively investigated because a large amount of laminin-111 can easily be prepared from mouse EHS tumor tissue. From the analysis of unglycosylated laminin-111, which is prepared from cell lysates of tunicamycin-treated cells, *N*-glycosylation of laminin-111 was revealed to be important for cell spreading and neurite outgrowth activities but not for cell adhesion activity or heterotrimer assembly [112]. However, because tunicamycin-treatment markedly inhibited secretion of laminin into the cell culture medium and unglycosylated laminin-111 was purified from cell lysates, the activities of unglycosylated laminin-111 purified from cell lysates may not be the same as that of the secreted protein. Therefore, the significance of *N*-glycosylation in laminin molecules remained unclear for a long time.

Recent studies using laminin-332 purified from the conditioned medium of glycosyltransferase-transfected cells clearly demonstrated that *N*-glycosylation of the laminin-332 molecule affects its biological functions, such as cell adhesion, migration, and scattering [113]. In some cancers, an increase of β1,6 GlcNAc, catalyzed by *N*-acetylglucosaminyltransferase V (GnT-V), is related to cancer metastasis, whereas bisecting GlcNAc suppresses further processing by branching enzymes, such as GnT-V, resulting in downregulation of cancer metastasis. Laminin-332 purified from the gastric cancer cell line MKN45 possesses a large amount of β1,6 GlcNAc and promptly induces cell adhesion and migration. In contrast, cell adhesion and migration activities of purified laminin-332 from GnT-III overexpressing MKN45 cells was apparently decreased compared with those of control laminin-332 purified from MKN45 cells. These weakened activities were most likely derived from the impaired α3β1 integrin clustering and resultant focal adhesion formation. Thus *N*-glycosylation of laminin-332 is important for its association with integrins and the subsequent cellular signaling (Figure 4).

Integrin is also a major carrier of *N*-glycans, and proper *N*-glycosylation of integrin is important for its functional activities and for heterodimer formation. A patient with a missense mutation in *ITGA3*, which introduces an additional *N*-glycosylation motif in the integrin α3 subunit, causes fatal interstitial lung disease and congenital nephrotic syndrome. The resulting gain of glycosylation impedes heterodimerization of the α3 precursor with β1 [114]. Recent studies using molecular and cellular biology-based techniques clearly show that *N*-glycans of both integrins α3β1 and α6β4 affect various cellular functions. Introduction of bisecting GlcNAc [115] or core fucose (α1,6-fucose) [116] into integrin α3β1 in MKN45 cells or α1,6-fucosyltransferase (Fut8)-null embryonic fibroblasts downregulates cell adhesion and migration through laminin-332. In contrast, the amount of β1,6 GlcNAc branched structures on integrin α3β1 correlated with metastatic ability in melanoma cells [117]. Although the precise mechanisms are unclear, *N*-glycosylation of integrin α3β1 is modulated by tetraspanin CD151 [118]. Furthermore, an *N*-glycosylation–defective CD151 mutant did not affect interactions of CD151 with other tetraspanins or integrin α3β1 but negated its modulatory function [118]. Galectin-3–mediated association among laminin-332, integrins, and EGFR through *N*-glycans is important for their complex formation and the subsequent cellular signaling. Galectins are a family of lectins, which bind β-galactoside and are often involved in protein–protein interactions by cross-linking β-galactosides on proteins through a carbohydrate recognition domain in the galectin molecule [119]. Galectin-3 is one of the best-characterized galectins, and it is only chimeric galectin [120]. Galectin-3 is ubiquitously expressed in adult tissues but is also highly expressed in some cancers as well as in wound healing. Binding of galectin-3 to β-galactoside on integrin α3β1, α6β4, and laminin-332 supports laminin-332-induced integrin clustering and focal contact formation [71, 113]. Galectin-3–mediated association between integrin α6β4 and EGFR is involved in integrin α6β4–EGFR complex formation, which activates ERK signaling pathway [71]. These galectin-3 mediated events are canceled by introduction of bisecting GlcNAc into these interacting proteins because bisecting GlcNAc inhibits GlcNAc branch formation catalyzed by GnT-IV and GnT-V, thereby suppressing addition of poly-*N*-acetyllactosamine *N*-glycan containing β-galactoside [71] (Figure 4). In the case of the integrin β4 subunit, expression of *N*-glycosylation-defective β4 integrin, in which asparagine (N) was replaced by glutamine (Q) in all five potential *N*-glycosylation sites of β4, decreased cell spreading, adhesion, migration, and lipid raft localization on laminin-332 in β4-null keratinocytes compared with β4-null keratinocytes expressing wild type β4 integrin [121]. Similar to the effect of *N*-glycans in CD151 on *N*-glycosylation of α3β1 [118], an *N*-glycosylation-defect in β4 integrin affects the *N*-glycosylation state of EGFR, which associates with β4 integrin through cross-linking with galectin-3–mediated *N*-glycans as well as by protein–protein interaction [121]. However, *N*-glycosylation-defective β4 integrin induced a strong association with EGFR, which is mediated only by protein–protein interaction but not by galectin-3–mediated *N*-glycan interaction compared with wild type β4 integrin [121]. Thus appropriate complex formation mediated by *N*-glycan, which would generate the adequate space between associated proteins, may have a suppressive effect on excessive protein–protein interactions and induce efficient cellular signaling and the subsequent cellular functions.

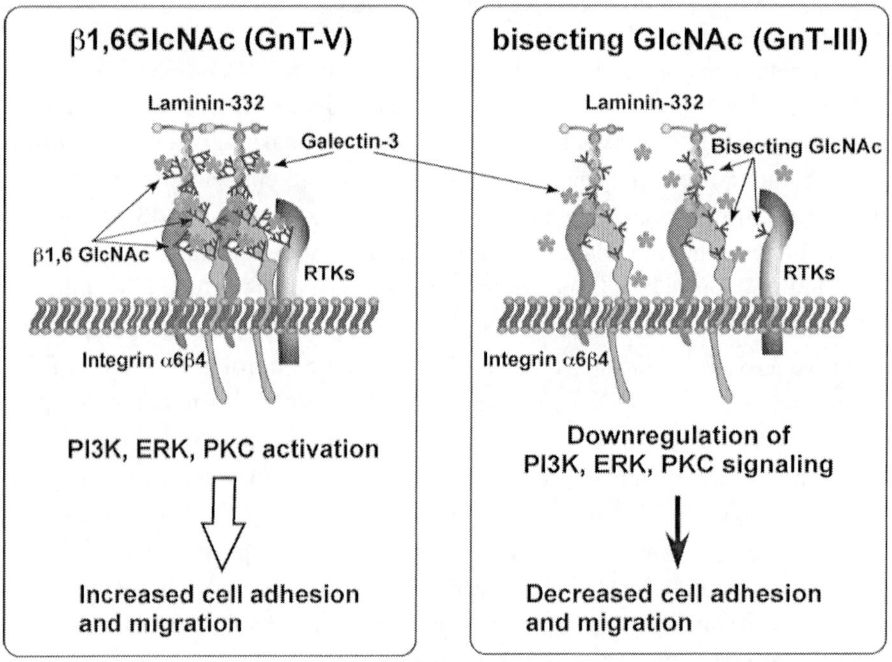

Figure 4. Model for the effect of *N*-glycans on laminin-332 and integrin α6β4-mediated signal transduction and cellular functions. With enhanced expression of GnT-V in cells (left panel), galectin-3 binds to integrin α6β4, laminin-332, and receptor tyrosine kinases (RTKs), such as EGFR, to form a complex on the cell surface responsible for cellular signaling, and cell adhesion and migration in cancer cells or normal cells such as keratinocytes. They usually express high or moderate levels of GnT-V and thus contain some poly-*N*-acetyllactosamine *N*-glycans. In this case, a β1,6GlcNAc–mediated cell signaling platform activates PI3K, ERK, and PKC signaling, thereby increasing cell adhesion and migration. In contrast, overexpression of GnT-III (right panel) results in modification of glycoproteins by bisecting GlcNAc, which inhibits GlcNAc branch formation catalyzed by GnT-V, thereby suppressing addition of poly-*N*-acetyllactosamine *N*-glycans to these proteins. Consequently, galectin-3 cannot form the signaling platform consisting of molecules such as RTKs, laminin-332, and α6β4 integrins, thereby inhibiting both cellular signaling and cellular function.

Laminin-332 in Stem Cell Biology

Human embryonic stem cells and induced pluripotent stem (iPS) cells have great potential for many applications in cellular regenerative therapy. Maintenance of these cells requires culture plate coatings. Although Matrigel is generally used as a culture substrate, its usage poses a problem for medical applications because most Matrigel components are undefined and it is prepared from mouse EHS tumors, and is not of human origin. Therefore, alternative substrates are required for culture systems. Recently, some laminin isoforms, including laminin-332, have become available as recombinant proteins [7, 122-124] and some studies have revealed that use of these recombinant laminin proteins can solve the above problems [125-129]. Further studies on the effects of recombinant laminins on cellular signaling of stem cells would not only clarify their cellular functions but would also help to establish an efficient culture system for these cells.

Conclusion

It has been more than 20 years since laminin-332 and its integrin receptors have been identified. Recent advances in investigations of these molecules provide us with new information to understand BMs and various diseases including epidermolysis bullosa and cancer. In both physiological and pathological conditions, laminin-332 and integrins associate with various binding proteins and cellular signaling molecules, and orchestrate them into a cellular signaling platform. The organized platform promotes cell adhesion, migration, and proliferation through the induction of various cellular signaling pathways, such as those of PI3K, ERK, PKC, JNK, and NFκB. The molecular constitution of the laminin-332/integrins-mediated signaling platform is altered by the proteolytic processing and *N*-glycosylation of laminin-332 as well as the phosphorylation or *N*-glycosylation of integrins. Further studies about this signaling platform are still needed to gain understanding of cell adhesion and migration under both physiological and pathological conditions. Characterization of the signaling processes mediated through laminin-332 and its associated integrins would help to design and develop new treatments and drugs for diseases as well as advance cellular therapy.

Acknowledgments

This work was partly supported by a Grant-in-Aid for the Uehara Memorial Foundation of Japan, and by a Grant-in-Aid for Young Scientists (B) No. 23790373 (Yo K), Scientific Research (21370059), for Challenging Exploratory Research (23651196) from the Japan Society for the Promotion of Science; by a grant from Scientific Research on Innovative Areas (23110002), and the Strategic Research Foundation Grant-Aided Project for Private Universities from the Ministry of Education, Culture, Sports, Science, and Technology of Japan as well as by the Mizutani Foundation for Glycoscience (JG).

References

[1] Kalluri, R. (2003). Basement membranes: structure, assembly and role in tumour angiogenesis. *Nat Rev Cancer, 3*, 422-433.

[2] Colognato, H. & Yurchenco, P. D. (2000). Form and function: the laminin family of heterotrimers. *Dev Dyn., 218*, 213-234.

[3] Domogatskaya, A., Rodin, S. & Tryggvason, K. (2012). Functional diversity of laminins. *Annu Rev Cell Dev Biol., 28*, 523-553.

[4] Miner, J. H., Patton, B. L., Lentz, S. I., Gilbert, D. J., Snider, W. D., Jenkins, N. A., Copeland, N. G. & Sanes, J. R. (1997). The laminin α chains: expression, developmental transitions, and chromosomal locations of α1-5, identification of heterotrimeric laminins 8-11, and cloning of a novel α3 isoform. *J Cell Biol., 137*, 685-701.

[5] Galliano, M. F., Aberdam, D., Aguzzi, A., Ortonne, J. P. & Meneguzzi, G. (1995). Cloning and complete primary structure of the mouse laminin α3 chain. Distinct expression pattern of the laminin α3A and α3B chain isoforms. *J Biol Chem., 270*, 21820-21826.

[6] Ryan, M. C., Tizard, R., VanDevanter, D. R. & Carter, W. G. (1994). Cloning of the LamA3 gene encoding the α3 chain of the adhesive ligand epiligrin. Expression in wound repair. *J Biol Chem., 269*, 22779-22787.

[7] Kariya, Y., Yasuda, C., Nakashima, Y., Ishida, K., Tsubota, Y. & Miyazaki, K. (2004). Characterization of laminin 5B and NH_2-terminal proteolytic fragment of its α3B chain: promotion of cellular adhesion, migration, and proliferation. *J Biol Chem., 279*, 24774-24784.

[8] Marinkovich, M. P., Lunstrum, G. P., Keene, D. R. & Burgeson, R. E. (1992). The dermal-epidermal junction of human skin contains a novel laminin variant. *J Cell Biol., 119*, 695-703.

[9] Champliaud, M. F., Lunstrum, G. P., Rousselle, P., Nishiyama, T., Keene, D. R. & Burgeson, R. E. (1996). Human amnion contains a novel laminin variant, laminin 7, which like laminin 6, covalently associates with laminin 5 to promote stable epithelial-stromal attachment. *J Cell Biol., 132*, 1189-1198.

[10] Kariya, Y., Mori, T., Yasuda, C., Watanabe, N., Kaneko, Y., Nakashima, Y., Ogawa, T. & Miyazaki, K. (2008). Localization of laminin α3B chain in vascular and epithelial basement membranes of normal human tissues and its down-regulation in skin cancers. *J Mol Histol., 39*, 435-446.

[11] Mori, T., Ono, K., Kariya, Y., Ogawa, T., Higashi, S. & Miyazaki, K. (2010). Laminin-3B11, a novel vascular-type laminin capable of inducing prominent lamellipodial protrusions in microvascular endothelial cells. *J Biol Chem., 285*, 35068-35078.

[12] Mori, T., Kariya, Y., Komiya, E., Higashi, S., Miyagi, Y., Sekiguchi, K. & Miyazaki, K. (2011). Downregulation of a newly identified laminin, laminin-3B11, in vascular basement membranes of invasive human breast cancers. *Cancer Sci., 102*, 1095-1100.

[13] Ryan, M. C., Lee, K., Miyashita, Y. & Carter, W. G. (1999). Targeted disruption of the LAMA3 gene in mice reveals abnormalities in survival and late stage differentiation of epithelial cells. *J Cell Biol., 145*, 1309-1323.

[14] Kuster, J. E., Guarnieri, M. H., Ault, J. G., Flaherty, L. & Swiatek, P. J. (1997). IAP insertion in the murine LamB3 gene results in junctional epidermolysis bullosa. *Mamm Genome., 8*, 673-681.

[15] Meng, X., Klement, J. F., Leperi, D. A., Birk, D. E., Sasaki, T., Timpl, R., Uitto, J. & Pulkkinen, L. (2003). Targeted inactivation of murine laminin γ2 chain gene recapitulates human junctional epidermolysis bullosa. *J Invest Dermatol., 121*, 720-731.

[16] Pulkkinen, L., Christiano, A. M., Gerecke, D., Wagman, D. W., Burgeson, R. E., Pittelkow, M. R. & Uitto, J. (1994). A homozygous nonsense mutation in the β3 chain gene of laminin 5 (LAMB3) in Herlitz junctional epidermolysis bullosa. *Genomics., 24*, 357-360.

[17] Aberdam, D., Galliano, M. F., Vailly, J., Pulkkinen, L., Bonifas, J., Christiano, A. M., Tryggvason, K., Uitto, J., Epstein, E. H., Jr., Ortonne, J. P. & Meneguzzi, G. (1994). Herlitz's junctional epidermolysis bullosa is linked to mutations in the gene (LAMC2) for the γ2 subunit of nicein/kalinin (LAMININ-5). *Nat Genet., 6*, 299-304.

[18] Varki, R., Sadowski, S., Pfendner, E. & Uitto, J. (2006). Epidermolysis bullosa. I. Molecular genetics of the junctional and hemidesmosomal variants. *J Med Genet., 43*, 641-652.

[19] Schneider, H., Muhle, C. & Pacho, F. (2007). Biological function of laminin-5 and pathogenic impact of its deficiency. *Eur J Cell Biol., 86*, 701-717.

[20] Sonnenberg, A., Calafat, J., Janssen, H., Daams, H., van der Raaij-Helmer, L. M., Falcioni, R., Kennel, S. J., Aplin, J. D., Baker, J., Loizidou, M. & Garrod, D. (1991). Integrin α6/β4 complex is located in hemidesmosomes, suggesting a major role in epidermal cell-basement membrane adhesion. *J Cell Biol., 113*, 907-917.

[21] Carter, W. G., Kaur, P., Gil, S. G., Gahr, P. J. & Wayner, E. A. (1990). Distinct functions for integrins α3β1 in focal adhesions and α6β4/bullous pemphigoid antigen in a new stable anchoring contact (SAC) of keratinocytes: relation to hemidesmosomes. *J Cell Biol., 111*, 3141-3154.

[22] Kariya, Y., Tsubota, Y., Hirosaki, T., Mizushima, H., Puzon-McLaughlin, W., Takada, Y. & Miyazaki, K. (2003). Differential regulation of cellular adhesion and migration by recombinant laminin-5 forms with partial deletion or mutation within the G3 domain of α3 chain. *J Cell Biochem., 88*, 506-520.

[23] Chen, M., Marinkovich, M. P., Jones, J. C., O'Toole, E. A., Li, Y. Y. & Woodley, D. T. (1999). NC1 domain of type VII collagen binds to the β3 chain of laminin 5 via a unique subdomain within the fibronectin-like repeats. *J Invest Dermatol., 112*, 177-183.

[24] Nakashima, Y., Kariya, Y., Yasuda, C. & Miyazaki, K. (2005). Regulation of cell adhesion and type VII collagen binding by the β3 chain short arm of laminin-5: effect of its proteolytic cleavage. *J Biochem., 138*, 539-552.

[25] Pulkkinen, L., Rouan, F., Bruckner-Tuderman, L., Wallerstein, R., Garzon, M., Brown, T., Smith, L., Carter, W. & Uitto, J. (1998). Novel ITGB4 mutations in lethal and nonlethal variants of epidermolysis bullosa with pyloric atresia: missense versus nonsense. *Am J Hum Genet., 63*, 1376-1387.

[26] van der Neut, R., Krimpenfort, P., Calafat, J., Niessen, C. M. & Sonnenberg, A. (1996). Epithelial detachment due to absence of hemidesmosomes in integrin β4 null mice. *Nat Genet., 13*, 366-369.

[27] Varki, R., Sadowski, S., Uitto, J. & Pfendner, E. (2007). Epidermolysis bullosa. II. Type VII collagen mutations and phenotype-genotype correlations in the dystrophic subtypes. *J Med Genet., 44*, 181-192.

[28] Ko, M.S. and Marinkovich, M.P. (2010). Role of dermal-epidermal basement membrane zone in skin, cancer, and developmental disorders. *Dermatol Clin., 28*, 1-16.

[29] Kariya, Y., Sato, H., Katou, N., Kariya, Y. & Miyazaki, K. (2012). Polymerized laminin-332 matrix supports rapid and tight adhesion of keratinocytes, suppressing cell migration. *PLoS One., 7*, e35546.

[30] Nguyen, B. P., Ryan, M. C., Gil, S. G. & Carter, W. G. (2000). Deposition of laminin 5 in epidermal wounds regulates integrin signaling and adhesion. *Curr Opin Cell Biol., 12*, 554-562.

[31] Sigle, R. O., Gil, S. G., Bhattacharya, M., Ryan, M. C., Yang, T. M., Brown, T. A., Boutaud, A., Miyashita, Y., Olerud, J. & Carter, W. G. (2004). Globular domains 4/5 of the laminin α3 chain mediate deposition of precursor laminin 5. *J Cell Sci., 117*, 4481-4494.

[32] Tsubota, Y., Yasuda, C., Kariya, Y., Ogawa, T., Hirosaki, T., Mizushima, H. & Miyazaki, K. (2005). Regulation of biological activity and matrix assembly of laminin-5 by COOH-terminal, LG4-5 domain of α3 chain. *J Biol Chem., 280*, 14370-14377.

[33] Gagnoux-Palacios, L., Allegra, M., Spirito, F., Pommeret, O., Romero, C., Ortonne, J. P. & Meneguzzi, G. (2001). The short arm of the laminin γ2 chain plays a pivotal role in the incorporation of laminin 5 into the extracellular matrix and in cell adhesion. *J Cell Biol., 153*, 835-850.

[34] Nakashima, Y., Kariya, Y. & Miyazaki, K. (2007). The β3 chain short arm of laminin-332 (laminin-5) induces matrix assembly and cell adhesion activity of laminin-511 (laminin-10). *J Cell Biochem., 100*, 545-556.

[35] Iriyama, S., Tsunenaga, M., Amano, S. & Adachi, E. (2011). Key role of heparan sulfate chains in assembly of anchoring complex at the dermal-epidermal junction. *Exp Dermatol., 20*, 953-955.

[36] Utani, A., Nomizu, M., Matsuura, H., Kato, K., Kobayashi, T., Takeda, U., Aota, S., Nielsen, P. K. & Shinkai, H. (2001). A unique sequence of the laminin α3 G domain binds to heparin and promotes cell adhesion through syndecan-2 and -4. *J Biol Chem., 276*, 28779-28788.

[37] Okamoto, O., Bachy, S., Odenthal, U., Bernaud, J., Rigal, D., Lortat-Jacob, H., Smyth, N. & Rousselle, P. (2003). Normal human keratinocytes bind to the α3LG4/5 domain of unprocessed laminin-5 through the receptor syndecan-1. *J Biol Chem., 278*, 44168-44177.

[38] Ogawa, T., Tsubota, Y., Hashimoto, J., Kariya, Y. & Miyazaki, K. (2007). The short arm of laminin γ2 chain of laminin-5 (laminin-332) binds syndecan-1 and regulates cellular adhesion and migration by suppressing phosphorylation of integrin β4 chain. *Mol Biol Cell., 18*, 1621-1633.

[39] Giancotti, F.G. and Tarone, G. (2003). Positional control of cell fate through joint integrin/receptor protein kinase signaling. *Annu Rev Cell Dev Biol., 19*, 173-206.

[40] Hynes, R.O. (2002). Integrins: bidirectional, allosteric signaling machines. *Cell., 110*, 673-687.

[41] Hemler, M.E. (2008). Targeting of tetraspanin proteins-potential benefits and strategies. *Nat Rev Drug Discov., 7*, 747-758.

[42] Lammerding, J., Kazarov, A. R., Huang, H., Lee, R. T. & Hemler, M. E. (2003). Tetraspanin CD151 regulates α6β1 integrin adhesion strengthening. *Proc Natl Acad Sci U S A., 100*, 7616-7621.

[43] Kazarov, A. R., Yang, X., Stipp, C. S., Sehgal, B. & Hemler, M. E. (2002). An extracellular site on tetraspanin CD151 determines α3 and α6 integrin-dependent cellular morphology. *J Cell Biol., 158*, 1299-1309.

[44] Niessen, C. M., Hogervorst, F., Jaspars, L. H., de Melker, A. A., Delwel, G. O., Hulsman, E. H., Kuikman, I. & Sonnenberg, A. (1994). The α6 β4 integrin is a receptor for both laminin and kalinin. *Exp Cell Res., 211*, 360-367.

[45] Carter, W. G., Ryan, M. C. & Gahr, P. J. (1991). Epiligrin, a new cell adhesion ligand for integrin α3β1 in epithelial basement membranes. *Cell., 65*, 599-610.

[46] Rousselle, P. and Aumailley, M. (1994). Kalinin is more efficient than laminin in promoting adhesion of primary keratinocytes and some other epithelial cells and has a different requirement for integrin receptors. *J Cell Biol., 125*, 205-214.

[47] Kikkawa, Y., Umeda, M. & Miyazaki, K. (1994). Marked stimulation of cell adhesion and motility by ladsin, a laminin-like scatter factor. *J Biochem., 116*, 862-869.

[48] Ido, H., Nakamura, A., Kobayashi, R., Ito, S., Li, S., Futaki, S. & Sekiguchi, K. (2007). The requirement of the glutamic acid residue at the third position from the carboxyl termini of the laminin γ chains in integrin binding by laminins. *J Biol Chem., 282*, 11144-11154.

[49] Kligys, K. R., Wu, Y., Hopkinson, S. B., Kaur, S., Platanias, L. C. & Jones, J. C. (2012). α6β4 integrin, a master regulator of expression of integrins in human keratinocytes. *J Biol Chem., 287*, 17975-17984.

[50] Choma, D. P., Pumiglia, K. & DiPersio, C. M. (2004). Integrin α3β1 directs the stabilization of a polarized lamellipodium in epithelial cells through activation of Rac1. *J Cell Sci., 117*, 3947-3959.

[51] Enserink, J. M., Price, L. S., Methi, T., Mahic, M., Sonnenberg, A., Bos, J. L. & Tasken, K. (2004). The cAMP-Epac-Rap1 pathway regulates cell spreading and cell adhesion to laminin-5 through the α3β1 integrin but not the α6β4 integrin. *J Biol Chem., 279*, 44889-44896.

[52] Has, C., Sparta, G., Kiritsi, D., Weibel, L., Moeller, A., Vega-Warner, V., Waters, A., He, Y., Anikster, Y., Esser, P., Straub, B. K., Hausser, I., Bockenhauer, D., Dekel, B., Hildebrandt, F., Bruckner-Tuderman, L. & Laube, G. F. (2012). Integrin α3 mutations with kidney, lung, and skin disease. *N Engl J Med., 366*, 1508-1514.

[53] DiPersio, C. M., Hodivala-Dilke, K. M., Jaenisch, R., Kreidberg, J. A. & Hynes, R. O. (1997). α3β1 Integrin is required for normal development of the epidermal basement membrane. *J Cell Biol., 137*, 729-742.

[54] Nakamura, K., Iwamoto, R. & Mekada, E. (1995). Membrane-anchored heparin-binding EGF-like growth factor (HB-EGF) and diphtheria toxin receptor-associated protein (DRAP27)/CD9 form a complex with integrin α3β1 at cell-cell contact sites. *J Cell Biol., 129*, 1691-1705.

[55] Wang, Z., Symons, J. M., Goldstein, S. L., McDonald, A., Miner, J. H. & Kreidberg, J. A. (1999). α3β1 integrin regulates epithelial cytoskeletal organization. *J Cell Sci., 112*, 2925-2935.
[56] Chattopadhyay, N., Wang, Z., Ashman, L. K., Brady-Kalnay, S. M. & Kreidberg, J. A. (2003). α3β1 integrin-CD151, a component of the cadherin-catenin complex, regulates PTPmu expression and cell-cell adhesion. *J Cell Biol., 163*, 1351-1362.
[57] Lampe, P. D., Nguyen, B. P., Gil, S., Usui, M., Olerud, J., Takada, Y. & Carter, W. G. (1998). Cellular interaction of integrin α3β1 with laminin 5 promotes gap junctional communication. *J Cell Biol., 143*, 1735-1747.
[58] deHart, G. W., Healy, K. E. & Jones, J. C. (2003). The role of α3β1 integrin in determining the supramolecular organization of laminin-5 in the extracellular matrix of keratinocytes. *Exp Cell Res., 283*, 67-79.
[59] Hamelers, I. H., Olivo, C., Mertens, A. E., Pegtel, D. M., van der Kammen, R. A., Sonnenberg, A. & Collard, J. G. (2005). The Rac activator Tiam1 is required for α3β1-mediated laminin-5 deposition, cell spreading, and cell migration. *J Cell Biol., 171*, 871-881.
[60] Litjens, S. H., de Pereda, J. M. & Sonnenberg, A. (2006). Current insights into the formation and breakdown of hemidesmosomes. *Trends Cell Biol., 16*, 376-383.
[61] Nakano, A., Pulkkinen, L., Murrell, D., Rico, J., Lucky, A. W., Garzon, M., Stevens, C. A., Robertson, S., Pfendner, E. & Uitto, J. (2001). Epidermolysis bullosa with congenital pyloric atresia: novel mutations in the β4 integrin gene (ITGB4) and genotype/phenotype correlations. *Pediatr Res., 49*, 618-626.
[62] Koster, J., Kuikman, I., Kreft, M. & Sonnenberg, A. (2001). Two different mutations in the cytoplasmic domain of the integrin β4 subunit in nonlethal forms of epidermolysis bullosa prevent interaction of β4 with plectin. *J Invest Dermatol., 117*, 1405-1411.
[63] Koster, J., van Wilpe, S., Kuikman, I., Litjens, S. H. & Sonnenberg, A. (2004). Role of binding of plectin to the integrin β4 subunit in the assembly of hemidesmosomes. *Mol Biol Cell., 15*, 1211-1223.
[64] Miyazaki, K., Kikkawa, Y., Nakamura, A., Yasumitsu, H. & Umeda, M. (1993). A large cell-adhesive scatter factor secreted by human gastric carcinoma cells. *Proc Natl Acad Sci U S A., 90*, 11767-11771.
[65] Kariya, Y. and Miyazaki, K. (2004). The basement membrane protein laminin-5 acts as a soluble cell motility factor. *Exp Cell Res., 297*, 508-520.
[66] Miyazaki, K. (2006). Laminin-5 (laminin-332): Unique biological activity and role in tumor growth and invasion. *Cancer Sci., 97*, 91-98.
[67] Marinkovich, M.P. (2007). Tumour microenvironment: laminin 332 in squamous-cell carcinoma. *Nat Rev Cancer., 7*, 370-380.
[68] Pyke, C., Salo, S., Ralfkiaer, E., Romer, J., Dano, K. & Tryggvason, K. (1995). Laminin-5 is a marker of invading cancer cells in some human carcinomas and is coexpressed with the receptor for urokinase plasminogen activator in budding cancer cells in colon adenocarcinomas. *Cancer Res., 55*, 4132-4139.
[69] Lohi, J. (2001). Laminin-5 in the progression of carcinomas. *Int J Cancer., 94*, 763-767.
[70] Tsuji, T., Kawada, Y., Kai-Murozono, M., Komatsu, S., Han, S. A., Takeuchi, K., Mizushima, H., Miyazaki, K. & Irimura, T. (2002). Regulation of melanoma cell

migration and invasion by laminin-5 and α3β1 integrin (VLA-3). *Clin Exp Metastasis., 19*, 127-134.

[71] Kariya, Y., Kawamura, C., Tabei, T. & Gu, J. (2010). Bisecting GlcNAc residues on laminin-332 down-regulate galectin-3-dependent keratinocyte motility. *J Biol Chem., 285*, 3330-3340.

[72] Pullar, C. E., Baier, B. S., Kariya, Y., Russell, A. J., Horst, B. A., Marinkovich, M. P. & Isseroff, R. R. (2006). β4 integrin and epidermal growth factor coordinately regulate electric field-mediated directional migration via Rac1. *Mol Biol Cell., 17*, 4925-4935.

[73] Nguyen, B. P., Gil, S. G. & Carter, W. G. (2000). Deposition of laminin 5 by keratinocytes regulates integrin adhesion and signaling. *J Biol Chem., 275*, 31896-31907.

[74] Hintermann, E., Bilban, M., Sharabi, A. & Quaranta, V. (2001). Inhibitory role of α6β4-associated erbB-2 and phosphoinositide 3-kinase in keratinocyte haptotactic migration dependent on α3β1 integrin. *J Cell Biol., 153*, 465-478.

[75] Nikolopoulos, S. N., Blaikie, P., Yoshioka, T., Guo, W., Puri, C., Tacchetti, C. & Giancotti, F. G. (2005). Targeted deletion of the integrin β4 signaling domain suppresses laminin-5-dependent nuclear entry of mitogen-activated protein kinases and NF-κB, causing defects in epidermal growth and migration. *Mol Cell Biol., 25*, 6090-6102.

[76] Marinkovich, M. P., Lunstrum, G. P. & Burgeson, R. E. (1992). The anchoring filament protein kalinin is synthesized and secreted as a high molecular weight precursor. *J Biol Chem., 267*, 17900-17906.

[77] Tsubota, Y., Mizushima, H., Hirosaki, T., Higashi, S., Yasumitsu, H. & Miyazaki, K. (2000). Isolation and activity of proteolytic fragment of laminin-5 α3 chain. *Biochem Biophys Res Commun., 278*, 614-620.

[78] Goldfinger, L. E., Stack, M. S. & Jones, J. C. (1998). Processing of laminin-5 and its functional consequences: role of plasmin and tissue-type plasminogen activator. *J Cell Biol., 141*, 255-265.

[79] Amano, S., Scott, I. C., Takahara, K., Koch, M., Champliaud, M. F., Gerecke, D. R., Keene, D. R., Hudson, D. L., Nishiyama, T., Lee, S., Greenspan, D. S. & Burgeson, R. E. (2000). Bone morphogenetic protein 1 is an extracellular processing enzyme of the laminin 5 γ2 chain. *J Biol Chem., 275*, 22728-22735.

[80] Veitch, D. P., Nokelainen, P., McGowan, K. A., Nguyen, T. T., Nguyen, N. E., Stephenson, R., Pappano, W. N., Keene, D. R., Spong, S. M., Greenspan, D. S., Findell, P. R. & Marinkovich, M. P. (2003). Mammalian tolloid metalloproteinase, and not matrix metalloprotease 2 or membrane type 1 metalloprotease, processes laminin-5 in keratinocytes and skin. *J Biol Chem., 278*, 15661-15668.

[81] Giannelli, G., Falk-Marzillier, J., Schiraldi, O., Stetler-Stevenson, W. G. & Quaranta, V. (1997). Induction of cell migration by matrix metalloprotease-2 cleavage of laminin-5. *Science., 277*, 225-228.

[82] Koshikawa, N., Giannelli, G., Cirulli, V., Miyazaki, K. & Quaranta, V. (2000). Role of cell surface metalloprotease MT1-MMP in epithelial cell migration over laminin-5. *J Cell Biol., 148*, 615-624.

[83] Mydel, P., Shipley, J. M., Adair-Kirk, T. L., Kelley, D. G., Broekelmann, T. J., Mecham, R. P. & Senior, R. M. (2008). Neutrophil elastase cleaves laminin-332

(laminin-5) generating peptides that are chemotactic for neutrophils. *J Biol Chem., 283*, 9513-9522.

[84] Udayakumar, T. S., Chen, M. L., Bair, E. L., Von Bredow, D. C., Cress, A. E., Nagle, R. B. & Bowden, G. T. (2003). Membrane type-1-matrix metalloproteinase expressed by prostate carcinoma cells cleaves human laminin-5 β3 chain and induces cell migration. *Cancer Res., 63*, 2292-2299.

[85] Remy, L., Trespeuch, C., Bachy, S., Scoazec, J. Y. & Rousselle, P. (2006). Matrilysin 1 influences colon carcinoma cell migration by cleavage of the laminin-5 β3 chain. *Cancer Res., 66*, 11228-11237.

[86] Tripathi, M., Nandana, S., Yamashita, H., Ganesan, R., Kirchhofer, D. & Quaranta, V. (2008). Laminin-332 is a substrate for hepsin, a protease associated with prostate cancer progression. *J Biol Chem., 283*, 30576-30584.

[87] Ogawa, T., Tsubota, Y., Maeda, M., Kariya, Y. & Miyazaki, K. (2004). Regulation of biological activity of laminin-5 by proteolytic processing of γ2 chain. *J Cell Biochem., 92*, 701-714.

[88] Bachy, S., Letourneur, F. & Rousselle, P. (2008). Syndecan-1 interaction with the LG4/5 domain in laminin-332 is essential for keratinocyte migration. *J Cell Physiol., 214*, 238-249.

[89] Mayer, U., Poschl, E., Gerecke, D. R., Wagman, D. W., Burgeson, R. E. & Timpl, R. (1995). Low nidogen affinity of laminin-5 can be attributed to two serine residues in EGF-like motif gamma 2III4. *FEBS Lett., 365*, 129-132.

[90] Geuijen, C.A. and Sonnenberg, A. (2002). Dynamics of the α6β4 integrin in keratinocytes. *Mol Biol Cell., 13*, 3845-3858.

[91] Margadant, C., Frijns, E., Wilhelmsen, K. & Sonnenberg, A. (2008). Regulation of hemidesmosome disassembly by growth factor receptors. *Curr Opin Cell Biol., 20*, 589-596.

[92] Germain, E. C., Santos, T. M. & Rabinovitz, I. (2009). Phosphorylation of a novel site on the β4 integrin at the trailing edge of migrating cells promotes hemidesmosome disassembly. *Mol Biol Cell., 20*, 56-67.

[93] Frijns, E., Kuikman, I., Litjens, S., Raspe, M., Jalink, K., Ports, M., Wilhelmsen, K. & Sonnenberg, A. (2012). Phosphorylation of threonine 1736 in the C-terminal tail of integrin β4 contributes to hemidesmosome disassembly. *Mol Biol Cell., 23*, 1475-1485.

[94] Guo, W. & Giancotti, F.G. (2004). Integrin signalling during tumour progression. *Nat Rev Mol Cell Biol., 5*, 816-826.

[95] Kariya, Y., Kariya, Y. & Gu, J. (2009). Roles of laminin-332 and α6β4 integrin in tumor progression. *Mini Rev Med Chem., 9*, 1284-1291.

[96] Dajee, M., Lazarov, M., Zhang, J. Y., Cai, T., Green, C. L., Russell, A. J., Marinkovich, M. P., Tao, S., Lin, Q., Kubo, Y. & Khavari, P. A. (2003). NF-κB blockade and oncogenic Ras trigger invasive human epidermal neoplasia. *Nature., 421*, 639-643.

[97] Tran, M., Rousselle, P., Nokelainen, P., Tallapragada, S., Nguyen, N. T., Fincher, E. F. & Marinkovich, M. P. (2008). Targeting a tumor-specific laminin domain critical for human carcinogenesis. *Cancer Res., 68*, 2885-2894.

[98] Waterman, E. A., Sakai, N., Nguyen, N. T., Horst, B. A., Veitch, D. P., Dey, C. N., Ortiz-Urda, S., Khavari, P. A. & Marinkovich, M. P. (2007). A laminin-collagen

complex drives human epidermal carcinogenesis through phosphoinositol-3-kinase activation. *Cancer Res., 67*, 4264-4270.

[99] Falcioni, R., Kennel, S. J., Giacomini, P., Zupi, G. & Sacchi, A. (1986). Expression of tumor antigen correlated with metastatic potential of Lewis lung carcinoma and B16 melanoma clones in mice. *Cancer Res., 46*, 5772-5778.

[100] Kimmel, K.A. & Carey, T.E. (1986). Altered expression in squamous carcinoma cells of an orientation restricted epithelial antigen detected by monoclonal antibody A9. *Cancer Res., 46*, 3614-3623.

[101] Mercurio, A.M. & Rabinovitz, I. (2001). Towards a mechanistic understanding of tumor invasion-lessons from the α6β4 integrin. *Semin Cancer Biol., 11*, 129-141.

[102] Rabinovitz, I., Toker, A. & Mercurio, A. M. (1999). Protein kinase C-dependent mobilization of the α6β4 integrin from hemidesmosomes and its association with actin-rich cell protrusions drive the chemotactic migration of carcinoma cells. *J Cell Biol., 146*, 1147-1160.

[103] Shaw, L. M., Rabinovitz, I., Wang, H. H., Toker, A. & Mercurio, A. M. (1997). Activation of phosphoinositide 3-OH kinase by the α6β4 integrin promotes carcinoma invasion. *Cell., 91*, 949-960.

[104] Mainiero, F., Murgia, C., Wary, K. K., Curatola, A. M., Pepe, A., Blumemberg, M., Westwick, J. K., Der, C. J. & Giancotti, F. G. (1997). The coupling of α6β4 integrin to Ras-MAP kinase pathways mediated by Shc controls keratinocyte proliferation. *Embo J., 16*, 2365-2375.

[105] Bon, G., Folgiero, V., Bossi, G., Felicioni, L., Marchetti, A., Sacchi, A. & Falcioni, R. (2006). Loss of β4 integrin subunit reduces the tumorigenicity of MCF7 mammary cells and causes apoptosis upon hormone deprivation. *Clin Cancer Res., 12*, 3280-3287.

[106] Dutta, U. & Shaw, L.M. (2008). A key tyrosine (Y1494) in the β4 integrin regulates multiple signaling pathways important for tumor development and progression. *Cancer Res., 68*, 8779-8787.

[107] Gambaletta, D., Marchetti, A., Benedetti, L., Mercurio, A. M., Sacchi, A. & Falcioni, R. (2000). Cooperative signaling between α6β4 integrin and ErbB-2 receptor is required to promote phosphatidylinositol 3-kinase-dependent invasion. *J Biol Chem., 275*, 10604-10610.

[108] Mizutani, K., Kawano, S., Minami, A., Waseda, M., Ikeda, W. & Takai, Y. (2011). Interaction of nectin-like molecule 2 with integrin α6β4 and inhibition of disassembly of integrin α6β4 from hemidesmosomes. *J Biol Chem., 286*, 36667-36676.

[109] Giannelli, G., Bergamini, C., Fransvea, E., Sgarra, C. & Antonaci, S. (2005). Laminin-5 with transforming growth factor-β1 induces epithelial to mesenchymal transition in hepatocellular carcinoma. *Gastroenterology., 129*, 1375-1383.

[110] Kim, Y., Kugler, M. C., Wei, Y., Kim, K. K., Li, X., Brumwell, A. N. & Chapman, H. A. (2009). Integrin α3β1-dependent β-catenin phosphorylation links epithelial Smad signaling to cell contacts. *J Cell Biol., 184*, 309-322.

[111] Fujiwara, S., Shinkai, H., Deutzmann, R., Paulsson, M. & Timpl, R. (1988). Structure and distribution of *N*-linked oligosaccharide chains on various domains of mouse tumour laminin. *Biochem J., 252*, 453-461.

[112] Dean, J. W., 3rd, Chandrasekaran, S. & Tanzer, M. L. (1990). A biological role of the carbohydrate moieties of laminin. *J Biol Chem., 265*, 12553-12562.

[113] Kariya, Y., Kato, R., Itoh, S., Fukuda, T., Shibukawa, Y., Sanzen, N., Sekiguchi, K., Wada, Y., Kawasaki, N. & Gu, J. (2008). *N*-glycosylation of laminin-332 regulates its biological functions. A novel function of the bisecting GlcNAc. *J Biol Chem., 283*, 33036-33045.

[114] Nicolaou, N., Margadant, C., Kevelam, S. H., Lilien, M. R., Oosterveld, M. J., Kreft, M., van Eerde, A. M., Pfundt, R., Terhal, P. A., van der Zwaag, B., Nikkels, P. G., Sachs, N., Goldschmeding, R., Knoers, N. V., Renkema, K. Y. & Sonnenberg, A. (2012). Gain of glycosylation in integrin α3 causes lung disease and nephrotic syndrome. *J Clin Invest., 122*, 4375-4387.

[115] Zhao, Y., Nakagawa, T., Itoh, S., Inamori, K., Isaji, T., Kariya, Y., Kondo, A., Miyoshi, E., Miyazaki, K., Kawasaki, N., Taniguchi, N. & Gu, J. (2006). *N*-acetylglucosaminyltransferase III antagonizes the effect of *N*-acetylglucosaminyltransferase V on α3β1 integrin-mediated cell migration. *J Biol Chem., 281*, 32122-32130.

[116] Zhao, Y., Itoh, S., Wang, X., Isaji, T., Miyoshi, E., Kariya, Y., Miyazaki, K., Kawasaki, N., Taniguchi, N. & Gu, J. (2006). Deletion of core fucosylation on α3β1 integrin down-regulates its functions. *J Biol Chem., 281*, 38343-38350.

[117] Pochec, E., Litynska, A., Amoresano, A. & Casbarra, A. (2003). Glycosylation profile of integrin α3β1 changes with melanoma progression. *Biochim Biophys Acta., 1643*, 113-123.

[118] Baldwin, G., Novitskaya, V., Sadej, R., Pochec, E., Litynska, A., Hartmann, C., Williams, J., Ashman, L., Eble, J. A. & Berditchevski, F. (2008). Tetraspanin CD151 regulates glycosylation of α3β1 integrin. *J Biol Chem., 283*, 35445-35454.

[119] Rabinovich, G. A., Toscano, M. A., Jackson, S. S. & Vasta, G. R. (2007). Functions of cell surface galectin-glycoprotein lattices. *Curr Opin Struct Biol., 17*, 513-520.

[120] Dumic, J., Dabelic, S. & Flogel, M. (2006). Galectin-3: an open-ended story. *Biochim Biophys Acta., 1760*, 616-635.

[121] Kariya, Y. & Gu, J. (2011). *N*-glycosylation of β4 integrin controls the adhesion and motility of keratinocytes. *PLoS One., 6*, e27084.

[122] Kariya, Y., Ishida, K., Tsubota, Y., Nakashima, Y., Hirosaki, T., Ogawa, T. & Miyazaki, K. (2002). Efficient expression system of human recombinant laminin-5. *J Biochem., 132*, 607-612.

[123] Kortesmaa, J., Yurchenco, P. & Tryggvason, K. (2000). Recombinant laminin-8 (α4β1γ1). Production, purification, and interactions with integrins. *J Biol Chem., 275*, 14853-14859.

[124] Doi, M., Thyboll, J., Kortesmaa, J., Jansson, K., Iivanainen, A., Parvardeh, M., Timpl, R., Hedin, U., Swedenborg, J. & Tryggvason, K. (2002). Recombinant human laminin-10 (α5β1γ1). Production, purification, and migration-promoting activity on vascular endothelial cells. *J Biol Chem., 277*, 12741-12748.

[125] Hashimoto, J., Kariya, Y. & Miyazaki, K. (2006). Regulation of proliferation and chondrogenic differentiation of human mesenchymal stem cells by laminin-5 (laminin-332). *Stem Cells., 24*, 2346-2354.

[126] Miyazaki, T., Futaki, S., Hasegawa, K., Kawasaki, M., Sanzen, N., Hayashi, M., Kawase, E., Sekiguchi, K., Nakatsuji, N. & Suemori, H. (2008). Recombinant human

laminin isoforms can support the undifferentiated growth of human embryonic stem cells. *Biochem Biophys Res Commun., 375*, 27-32.

[127] Rodin, S., Domogatskaya, A., Strom, S., Hansson, E. M., Chien, K. R., Inzunza, J., Hovatta, O. & Tryggvason, K. (2010). Long-term self-renewal of human pluripotent stem cells on human recombinant laminin-511. *Nat Biotechnol., 28*, 611-615.

[128] Domogatskaya, A., Rodin, S., Boutaud, A. & Tryggvason, K. (2008). Laminin-511 but not -332, -111, or -411 enables mouse embryonic stem cell self-renewal in vitro. *Stem Cells., 26*, 2800-2809.

[129] Miyazaki, T., Futaki, S., Suemori, H., Taniguchi, Y., Yamada, M., Kawasaki, M., Hayashi, M., Kumagai, H., Nakatsuji, N., Sekiguchi, K. & Kawase, E. (2012). Laminin E8 fragments support efficient adhesion and expansion of dissociated human pluripotent stem cells. *Nat Commun., 3*, 1236., 10.1038/ncomms2231.

In: Laminins: Structure, Biological Activity and Role in Disease ISBN: 978-1-62618-568-5
Editors: Demetrius C. Adams and Emil O. Garcia

Chapter 3

Anosmin-1 and Galectin-1 as Laminin Interactors in Nervous System Development and Pathology

Silvia Velasco[1] and Fernando de Castro[2]
[1]Grupo de Biología de Membrana y Reparación Axonal,
Hospital Nacional de Parapléjicos-SESCAM, Toledo, Spain
[2]Grupo de Neurobiología del Desarrollo-GNDe,
Hospital Nacional de Parapléjicos-SESCAM, Toledo, Spain

Abstract

The laminin family is one of the most studied adhesion molecules. Present in multiple tissues during development and adulthood, in physiological and pathological conditions, laminins are capital for cell adhesion and consequently have been shown to play roles in cell survival, migration and differentiation. Particularly in the nervous system, during development laminin is fundamental for the adhesion of neurons, astrocytes and oligodendrocytes, and, as consequence, takes part in the control of the migration of all these cell types as well as in the outgrowth of axons. In the last years, different molecules have been identified as interacting with laminin and two of them, the glycoprotein anosmin-1 and the β-galactoside-binding protein galectin-1, have been shown to be involved in neuronal, astroglial, oligodendroglial development as well as in different neurological diseases. In the present work we describe the mechanisms used by anosmin-1 and galectin-1 to bind laminin, and we review the relative relevance of these in the bulk of biological roles that both proteins have been demonstrated to play during nervous system development and in pathology.

Introduction

Laminin glycoproteins are major constituents of basement membranes (basal lamina), extracellular matrices (ECMs) in intimate contact with cells and cell layers. Through the association with macromolecules of the ECM and interaction with integrins, proteoglycans and specific cell surface molecules, laminins are involved in cell survival, adhesion, migration, and differentiation (including axogenesis and axon guidance), playing crucial roles in physiological and pathological processes (Colognato and Yurchenco, 2000). The assembly and supramolecular architecture of basement membrane is a prerequisite for proper nervous system development (Garcia-Alonso et al., 1996; Lallier et al., 1996). Indeed the basal lamina surrounds the brain and blood vessels throughout the CNS, critically contributing to form the blood-brain barrier as well as lamination (Santos and Müller, 2011). Furthermore, laminin exhibits a variety of other functions, such as the ability to promote neurite outgrowth, migration, growth cone guidance, pathfinding and synapse formation (Venstrom and Reichardt, 1993; Luckenbill-Edds, 1997). Most of these processes involve the specific interaction of laminin with ligands expressed in the ECM or on the surface of neural and glial cells, like anosmin-1 and galectin-1.

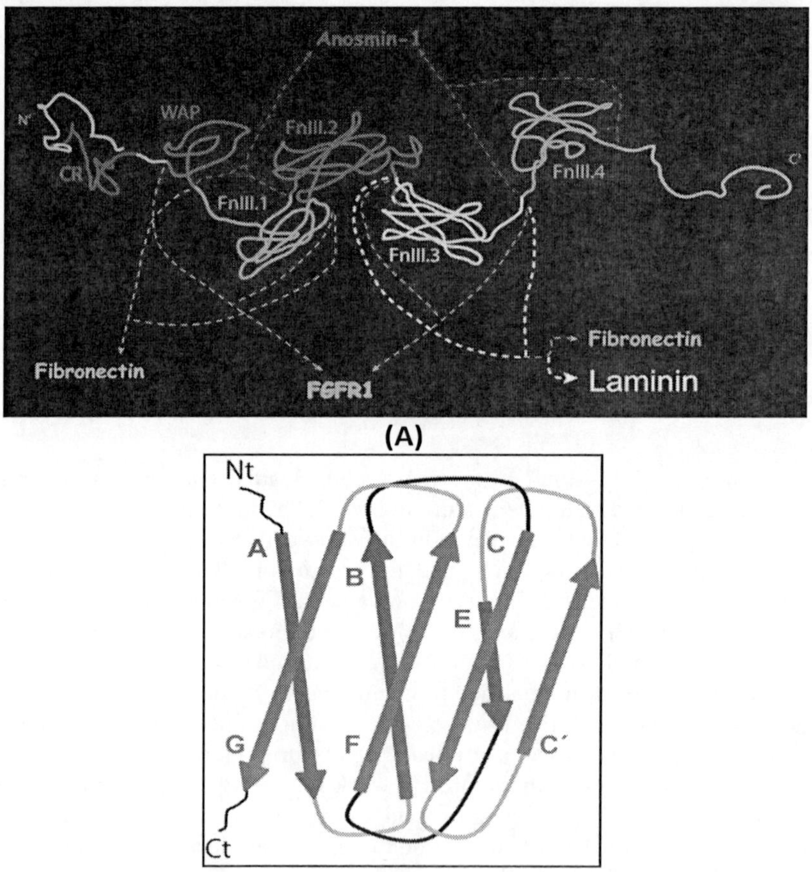

Figure 1. Anosmin-1 sequence and interaction with laminin and other components of the ECM. A: Schematic representation of anosmin-1 sequence and its interaction map. Its binding to laminin is only due to the FnIII.3 domain. This is also used by fibronectin together with FnIII.1. All laminin, fibronectin and anosmin-1 itself are ECM proteins, while FGFR1 is a receptor on the cell surface. B: Detailed schematic representation of the structure of FnIII.3, the only anosmin-1 domain involved in laminin-anosmin-1 interaction. A has been adapted from Murcia-Belmonte et al. 2010.

Anosmin-1 is encoded by *Kal-1* gene and was identified as defective in human Kallmann syndrome (Franco et al., 1991; Legouis et al., 1991). This 680 aminoacid glycoprotein is secreted to the ECM. Different domains can be distinguished in its structure: a cysteine-rich region (CR domain), followed by a whey acidic protein-like domain (WAP) similar to that found in serin-protease inhibitors, then four consecutive fibronectin type III domains (FnIII), and a C-terminal region rich in histidines and prolines (del Castillo et al., 1992; Legouis et al., 1993; Rugarli et al., 1996). This protein sequence shows up to five potential heparan sulphate-binding motifs which determine anosmin-1 localisation in ECM (Robertson et al., 2001), as well as up to six putative sites for N-glycosilation with not identified functionality for the protein (Rugarli et al., 1996). Thanks to these disposition and composition, anosmin-1 presents important basic-charged surfaces within the FnIII domains that define the binding to the different interacting molecules (Robertson et al., 2001) (Figure 1A and 1B).

Figure 2. Schematic representation of a functional Galectin-1 homodimer. 3D ribbon diagram of the homodimeric human Gal-1 protein. Red arrows indicate sites for carbohydrate recognition responsible for laminin-binding. Adapted from Lopez-Lucendo et al. (2004).

Galectin-1 (Gal-1) is a member of the family of galectins, highly conserved and developmentally regulated proteins, which share two common properties: i) affinity for β-galactosides and ii) significant sequence similarity in the carbohydrate recognition domain (CRD). Based on their domain organization, galectins are classified into three different groups. Prototype galectins are monomers containing one single CRD, chimera-type consist of one CRD linked to a N-terminal domain and, tandem-repeat galectins carry two distinct CRDs in the same peptide. Gal-1, the first prototype galectin described (Gitt et al., 1992), is a 14kDa-monomer, however it can form homodimers through non-covalent self-association (Figure 2). This property is foundamental for cross-linking cell surface ligands and thus for Gal-1 biological activity (Patnaik et al., 2006). Gal-1 shows widespread distribution and is concentrated in either the cytoplasm of the cell or in the ECM, depending on the tissue and the stage of development. As a consequence, Gal-1 is involved in either intracellular processes, such as pre-mRNA splicing, apoptosis and cell cycle regulation or in extracellular functions including cell-cell and cell-ECM adhesion, cell migration and intercellular interactions (Figure 3).

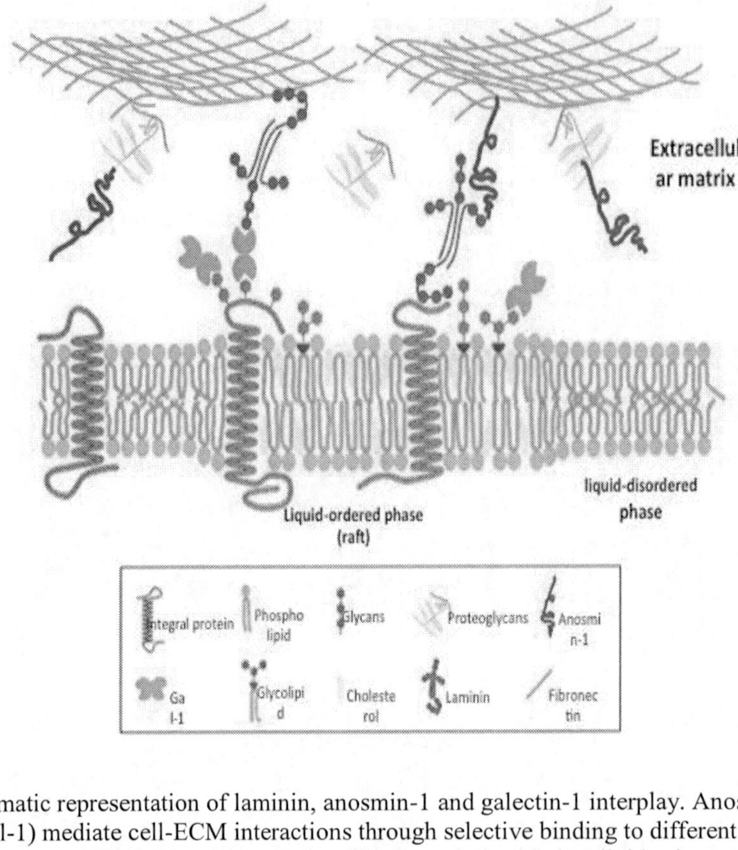

Figure 3. Schematic representation of laminin, anosmin-1 and galectin-1 interplay. Anosmin-1 and Galectin-1 (Gal-1) mediate cell-ECM interactions through selective binding to different ligands, such as laminin. Different transmembrane glycoproteins (like integrins) and glycolipids give support to Gal-1 in the surface of the cell. Laminin can bind, both directly and indirectly, to integrins. ECM anosmin-1 binds also laminin, but direct Gal-1/anosmin-1 binding has not reported to date. The plasmalemmal membrane has been adapted from Fijałkowski A.J. (http://commons.wikimedia.org/wiki/File:Lipid_raft_organisation_scheme.svg).

The Interaction between Anosmin-1 and Laminin

Anosmin-1 is known to interact with an important number of molecules of different nature: ECM components (urokinase-type plasminogen activator, fibronectin, laminin, anosmin-1 itself), maybe Prokineticine 2 and its receptor Prokr2, morphogens/growth factors (FGF8, BMP5, WNT3a) and, finally, the two more intensively studied to date, HS and FGFR1 (Robertson et al., 2001; González-Martínez et al., 2004; Hu et al., 2004; Bribián et al. 2006; Ayari and Soussi-Yanicostas 2007; Bribián et al., 2008; Hu et al., 2009; Murcia-Belmonte et al., 2010; Endo et al., 2012; Esteban et al., 2012; for a detailed review about these interactions, see: de Castro et al., 2013). Among ECM components (Bribián et al., 2008; Hardelin and Dodé, 2008; Murcia-Belmonte et al., 2010), anosmin-1 binding to laminin has been demonstrated to be selectively due to the FnIII.3 domain (Murcia-Belmonte et al., 2010)

(Figure 1A and 1B). Although the FnIII.3 domain is implicated in the binding of the protein to FGFR1 and fibronectin, too, the mutations found in FnIII.3 that impede anosmin-1 binding to FGFR1 have little or no effect in the binding to laminin or fibronectin (Murcia-Belmonte et al., 2010).

The Interaction between Galectin-1 and Laminin

Gal-1 shows high affinity for glycoconjugates carrying multiple N-acetyllactosamine sequences (Leffler and Barondes, 1986; Merkle and Cummings, 1988), like those present in laminin (Zhou and Cummings, 1990; Castronovo et al., 1992), lysosome-associated membrane proteins (Do et al., 1990; Skrincosky et al., 1993), a lactosamine containing glycolipid on olfactory neurons (Mahanthappa et al., 1994) and integrin $\alpha7\beta1$ on muscle cells (Gu et al., 1994). Trough interaction with the poly-N-acetyllactosamine chains on complex-type Asn-linked oligosaccharides of laminin, Gal-1 plays a role in the assembly of the ECM and in the control of cell-ECM adhesion (Figure 3). The effects of Gal-1 on cell-adhesion to ECM vary depending on tissue and developmental state. In skeletal muscle, for instance, Gal-1 is secreted during differentiation and binds specifically to laminin in basement membrane interfering with its association with integrin $\alpha7\beta1$. The inhibition of cell adhesion to laminin is critical for myoblast migration, alignment and fusion (Cooper et al., 1991). However in other tissues or in muscle at different stage of development, Gal-1 might promote cell adhesion through its bivalency and cross-linking ability, stabilizing the association between ECM and cell surface glycoconjugates.

Effects of Anosmin-1 in CNS Development

As a classical ECM protein, anosmin-1 plays relevant roles in substrate cell adhesion, migration and cell differentiation, including axonal outgrowth. Anosmin-1 is a chemoattractant for GnRH-expressing neurons and mutations in the protein cease this effect (Cariboni et al., 2004; Whitlock et al., 2005; Okubo et al., 2006). Together with FGF-2, anosmin-1 is also a chemoattractant for SVZ neuroblasts migrating towards the OB during pre- and postnatal development, before the rostral migratory stream acquires its mature aspect, although this effect is mainly mediated via FGFR1 (García-González et al., 2010; Murcia-Belmonte et al., 2010). In the zebrafish, the lack of anosmin-1 depletes the olfactory bulb from interneurons, generated at the SVZ (Yanicostas et al., 2009). Anosmin-1 seems crucial for the migration of the posterior lateral line primordium cells in fishes, although this effect seems to be clearly mediated by CXCRs/SDF1a signalling pathways (Yanicostas et al., 2008). The migration of neuroblasts in invertebrates is FGF-independent effect, mediated by heparan sulphate proteoglycans (Hudson et al., 2006). On the contrary, in oligodendrocyte precursors cells (OPCs), the balance between anosmin-1 and FGF-2 modulates OPC migration through their interaction with FGFR1, and physiologically anosmin-1-expressing OPCs adhere to laminin-, fibronectin- and anosmin-1-substrates, reducing cell motility (Bribián et al., 2006; Bribián et al., 2008). This latter effect is FGFR1-independent (Bribián et al., 2008). The correct OPC migration from oligodendrogenic sites to their final emplacement in the CNS is indispensable for the correct myelination of nerve fibres and, subsequently, the physiology of the CNS (de Castro and Bribián, 2005).

In human embryonic GnRH-derived cells, anosmin-1 induces neurite outgrowth and cytoskeletal rearrangements, both through FGFR1-dependent mechanisms (González-Martínez et al., 2004). In fishes, the lack of anosmin-1a impairs the fasciculation of olfactory axons and their projection onto the olfactory bulb (Yanicostas et al., 2009), while in mammals anosmin-1 attracts the growth of axons from olfactory bulb projection neurons (mitral and tufted cells) which form the lateral olfactory tract, enhances collateral branching of these axons and attracts their growth to invade the piriform cortex (Soussi-Yanicostas et al., 2002). Similar effects are observed in the Purkinje cells at the cerebellum: anosmin-1 promotes the outgrowth and fasciculation of their axons and strongly increases the formation of axon collaterals and the extension of terminal arbours, both at embryonic (E17) and newborn (P0) stages, as well as promotes Purkinje axons regeneration after axotomy (Gianola et al., 2009). The mechanism/s underlying anosmin-1 effects on the axons of these projection neuron remains to be elucidated.

Effects of Galectin-1 in CNS Development

Several studies show that Gal-1 participates in the development and maintenance of neural tissue function. Through binding to lactosamine-containing glycans expressed on neuronal surface and laminin in the ECM of the axonal path, Gal-1 plays a crucial role in inducing axonal fasciculation and in promoting axon growth and guidance, essential for neuronal pathfinding in the developing and adult olfactory nervous system (Mahanthappa et al., 1994). Indeed, in *Gal-1*-null mutant mice a subpopulation of olfactory neurons fail to correctly target the dorsocaudal region of the olfactory bulb (Puche et al., 1996). Gal-1 is also required for proper growth and guidance of subsets of primary somatosensory axons to their targets in the superficial dorsal horn of the spinal cord. Studies on *Gal-1* deficient mice confirm that Gal-1 is involved in peripheral nociceptive neuronal development and that its absence results in anatomical and functional deficits in adulthood (McGraw et al., 2005). Moreover, Gal-1 induces astrocyte differentiation and secretion of brain-derived neurotrophic factor (BDNF) and, as a consequence, promotes neuronal survival, axonal guidance and synaptic plasticity, essential for development and maintenance of nervous system (Sasaki et al., 2004). In addition to that, Gal-1 has recently been identified as a key molecule that promotes the proliferation of adult neural progenitor cells in the subventricular zone of the lateral ventricules (Sakaguchi et al., 2006; Sakaguchi et al., 2007) and in the subgranular zone of the hippocampal dentate gyrus (Kajitanie et al., 2009). The regulation of adult neural progenitor cells depends on mechanisms of cell adhesion involving the interaction of Gal-1 with β1 integrin (Sakaguchi et al., 2010). Considering that laminin is a known binding partner of both Gal-1 and β1-integrin subunit, it might participate in modulating this process. Galectin-glycan interactions play also a role in regulating oligodendrocyte differentiation leading to the control of myelin integrity and function. In contrast to Gal-3, Gal-1 appears to be an inhibitor of oligodendrocyte maturation. The functional differences in the biological activities of these proteins might be explained by considering that immature oligodendrocyte display intermediate levels of α2-6-linked sialic acid, which are permissive for Gal-3 binding, but are more resctrictive for Gal-1 recognition (Pasquini et al., 2011).

Effects of Anosmin-1 in CNS Pathology

Obviously, for most of the scientific community, anosmin-1 is a synonim for Kallmann syndrome (KS). This is a developmental disease characterised by hypogonadotropic hypogonadism and anosmia, with an estimated prevalence around 1:8,000 and 1:40,000 in women (Dodé and Hardelin, 2009). The mutat+ion of the gene encoding anosmin-1 is one the most common aethiologies for this syndrome, the X-linked form, but it is not the only one (Hardelin and Dodé, 2008). Defects in the sense of smell, either anosmia (complete) or at least severe hyposmia, combine with a deficiency in hypothalamic GnRH-1 cell function, which associates with different defects in gonads, causing infertility once puberty arrives. The first descriptor of KS was the Spanish pathologist Aureliano Maestre de San Juan (Maestre de San Juan, 1856), although the genetic link of the syndrome was reported almost a century later (Kallmann et al., 1944; De Morsier, 1954). The most common cause for hypogonadotropic hypogonadism in KS is the defective migration of GnRH-1 neurons, which do not properly integrate in the hypothalamic structures, and this results in improper GnRH secretion and the subsequent problem in peripheral gonads development; on the contrary, a variety of anatomical defects could be responsible for anosmia, ranging from defects in the formation of the olfactory nerve, agenesia/hypoplasia of the olfactory bulb or defects in the lateral and medial olfactory tracts, the central afferences to the olfactory cortex (Cariboni et al., 2004). Both symptoms should derive from defects in anosmin-1/*Kal-1* gene, with or without affection of FGF signalling (Chung and Tsai, 2010). Other gene, *KAL2* (which encodes FGFR1), is responsible for the autosomal dominant variant of KS (Dodé et al., 2003), and many KS satellite symptoms (loss of nasal cartilage, major hearing defects, abnormal extremities) are mainly associated with *KAL2* mutations.

In the last years, our group has described the up-regulation of anosmin-1 in part of the demyelinating lesions studied in samples of the brain of patients with multiple sclerosis (MS), the most frequent demyelinating disease in young adults (Clemente et al., 2011). MS lesions show demyelination, inflammation, axonal damage, reactive gliosis and, spontaneously, it would be observed a variable degree of remyelination, which can be extensive in a number of cases (Frohman et al., 2006; Patani et al., 2007; Patrikios et al., 2006). However, in chronic MS lesions inflammation and remyelination are absent in the core of the lesion and limited to the periplaque zone (Barkhof et al., 2003;Bramow et al., 2010; Prineas and Connell, 1979). Limited immigration of endogenous oligodendrocyte precursor cells, together with the blockade of the differentiation of this cell type determine spontaneous remyelination failure in chronic MS lesions (Dubois-Dalcq et al., 2005; Kuhlmann et al., 2008). Associated to demyelinated plaques, different molecules involved in oligodendrocyte precursor cell biology during development, like CXCL1, semaphorins 3A and 3F, and sonic hedgehog, are observed (de Castro et al., *Submitted*), and particularly relevant seems the case of FGF2 and anosmin-1, which show a complementary distribution in the different histopathological types of demyelinated plaques in MS patients (Clemente et al., 2011): in areas with active remyelinating activity, i.e. active lesions and the periplaque of chronic-active plaques, there is an infiltration of FGF-2$^+$-macrophages, whereas anosmin-1 is absent; in contrast, where the remyelination process is completely compromised i.e. demyelinated area of chronic-active and chronic-inactive plaques, FGF2 is totally absent while anosmin-1 fills the entire core of lesions. In the latter case, anosmin-1 may be synthesised by astrocytes, as observed in the cerebellum during development (Gianola et al., 2009). In addition, anosmin-1 is present in *approximately* 15 % of the nude axons that traverse the demyelinated area in chronic lesions, but not in the periplaque or in the adjacent normal appearing white matter (Clemente et al., 2011). This confirms that axons in the brain of adult MS patients adopt developmental features, re-expressing not only anosmin-1 but other ECM components like PSA-NCAM, glycosaminglycan hyaluronan or another glycoprotein like CD44 (Charles *et al*, 2002; Bribián et al., 2008; Soussi-Yanicostas et al., 2002; Back et al., 2005; Clemente et al., 2011).

Besides the above mentioned roles in neurologic diseases, *KAL-1* gene is up-regulated in breast, lung, ovarian and colon tumours (Jian et al., 2009; Arikawa et al., 2011), and is important for cancer metastasis and protects malignant cells from apoptosis (Jian et al., 2009). The correlation of up-regulated *KAL-1* gene with the cancer progression from early (I and II) to advanced stages (III and IV) is specially significant in colon cancer, and in related cell lines both TGF-β and hypoxia induce *KAL-1* gene expression and secretion of anosmin-1, which in turn protects cancer cells from hypoxia-induced apoptosis and increases the motility of the cells favouring their invasiveness (Jian et al., 2009). Finally, in atopic dermatitis anosmin-1 is produced by epidermal keratinocytes in response to increased calcium concentration or anti-inflammatory cytokines, maybe to control nerve density of epidermis (Tengara et al., 2010).

Effects of Galectin-1 in CNS Pathology

Gal-1 has been implicated in a variety of CNS disorders including neuroinflammation, neurodegeneration and nervous system tumorigenesis.

Galectins are important immunomodulators in diverse inflammatory diseases and they might also play a role in multiple sclerosis (MS) pathophysiology (Stancic et al., 2011). Indeed, the expression of several galectins is significantly increased in active MS lesions and the administration of Gal-1 and Gal-9, which are generally regarded as anti-inflammatory, suppress clinical and histological signs of experimental autoimmune encephalomyelitits (EAE), an animal model of MS (Offner et al., 1990; Zhu et al., 2005). Consistently, *Gal-1* knockout animals showed an enhanced susceptibility to develop EAE, and suppression of Gal-9 expression increases the clinical severity of EAE symptoms (Toscano et al., 2007). Moreover, approximately one third of the MS patient population shows high titres of anti-Gal-1 autoantibodies (Lutomski et al., 1997). Several findings suggest that Gal-1 is also involved in the pathogenesis of amiotrophic lateral sclerosis (ALS). Indeed, Gal-1 accumulates in the neurofilamentous lesions associated to motor neuron degeneration in sporadic and familial ALS (Kato et al., 2001). Furthermore a marked depletion of Gal-1 immunoreactivity has been observed in the skin of patients with ALS, another abnormality frequently observed in ALS (Wada et al., 2003). Of note, the administration of oxidized Gal-1 has neuroprotective effect in a transgenic mouse model of ALS (Chang-Hong et al., 2005; Kato et al., 2005). Moreover, the upregulation of Gal-1 is implicated in neuroprotection, neurogenesis (Kurushima et al., 2005; Ishibashi et al., 2007) and, as a whole, promotes recovery after brain ischemia (Qu WS 2011). An increase in Gal-1 expression has also been observed in immune cells and glial cells assembling around the lesion site in injured spinal cord (Kurihara et al., 2010). Remarkably it has been suggested that Gal-1 acts synergically with reactive astrocytes to enhance functional recovery after contusive spinal cord injury (SCI), showing a great potential as a novel neuroprotective agent for the treatment of this condition (Han et al., 2011). Considering its function in the somatosensory system, Gal-1 might also contribute to neuropathic pain. Furthermore Gal-1 plays a crucial role in the processes of nerve injury and regeneration in both CNS, where it induces the activation of ΔFosB trascription factor (Kurushima et al., 2005), and in PNS, stimulating macrophages to secrete a factor that promotes axonal growth and Schwann cell migration (Horie et al., 1999; Horie et al., 2004; Horie et al., 2005). Importantly, Gal-1 is able to induce axon regeneration only in its oxidized form, which lacks lectin activity. Considering the role of Gal-1 in adult hippocampal neurogenesis and that this process is related to learning and memory activities, it is likely that galectin-carbohydrate signaling also contribute to cognitive functions. Indeed *Gal-1* deficient mice show impaired spatial and contextual memory formation (Sakaguchi et al., 2011). Marked changes in the expression of specific galectins have also been described in various cancers (Danguy et al., 2002). Gal-1 is expressed in all human glioma types, including astrocytic, oligodendroglial and ependymal tumors and its expression correlates specifically with the grade of aggressiveness of astrocytic tumors (Rorive et al., 2001). Considering that Gal-1 enhances the migratory capabilities of tumor astrocytes and interact with laminin, it is tempting to suggest that it might be a potential modulator of tumor cell-laminin interactions, whose alteration has been observed in several invasive and metastatic tumors. Indeed there is evidence that Gal-1 regulate adhesion of melanoma cells to laminin and, likely, could be a modulator of invasion and metastasis even in other types of tumors, such as astrocytomas (Van de Brûle et al., 1995).

Apart from the role in promoting axon regeneration, which requires the oxidized form of Gal-1, lacking lectin properties, all the other effects of Gal-1 depends on its ability to recognize specific carbohydrate structures. Despite the precise mechanisms have not been completely elucidated, the interaction of Gal-1 with adhesion molecules like laminin might play an important role in some of these processes. Indeed, similarly to Gal-1, changes in laminin expression have been reported in various CNS pathological conditions like Alzheimer's disease, Down syndrome (Murtomäki et al., 1992) and brain ischemia (Liesi et al., 1984).

Conclusion

The interactions of cells with components of the extracellular environment has profound effects on developmental morphogenesis, cell differentiation, stabilization of intercellular contacts and physiology of the mature cells. There is an increasing bulk of evidences of all these in the CNS and molecules like the glycoprotein anosmin-1 and the β-galactoside-binding protein Gal-1 can specifically bind to laminin and other ECM components, playing an essential role in these processes. Although the mot studied mechanism of action for anosmin-1 is its binding to FGFR1 and subsequent modulation of cell signalling agonising or antagonising FGF activity, anosmin-1 binding to laminin and other components of ECM is also relevant for the function of the protein in physiology and pathology. Most of the extracellular biological functions of Gal-1 depends on its ability to interact with glycans, like those carried by laminin. Anosmin-1/ and Gal-1/laminin interactions should be increasingly considered in development and in adulthood, in physiological and pathological scenarios. Besides of the roles in KS, anosmin-1 has been recently identified as relevant in human MS. Although to date there are few reports on the role of Gal-1 in MS, evidences regarding its role in biology of oligodendroglia invite to perform a careful study in the future. Both, anosmin-1 and Gal-1 seem to be relevant signs of malignancy and metastatic progression of different cancers. On the other hand, Gal-1 in its oxidized form plays a number of important roles in the regeneration of both the CNS and the PNS after injury, aspects which remains to be studied for anosmin-1. Understanding the implication of both anosmin-1/ and Gal-1/laminin interactions in these physiological and pathological proceses might thus contribute to the comprehension of important biological processes and to the development of future neurorepair therapies and new ways to treat neoplastic processes.

Acknowledgments

Research at Grupo de Neurobiología del Desarrollo-GNDe is currently supported by grants from the Spanish Ministerio de Economía y Competitividad-MINECO (ADE10-0010, RD07-0060-2007, SAF2009-07842; SAF2012-40023), Fundación Eugenio Rodríguez Pascual (Spain), and Association pour la Recherche en Escleròse en Plaques-ARSEP (France) to FdC. FdCS is hired by SESCAM/Gobierno de Castilla-La Mancha.

Research at Grupo de Biología de Membrana y Reparación Axonal was funded by EC Marie Curie RTN (contract no. 2005-019561), EU 7th Framework Program (under grant agreement no. 26060), and Carlos III Health Institute and Castilla-La-Mancha Health Service (SESCAM) EMER program (EMER07/026). SV initially joined the DYNAMIC Marie Curie Research Training and Mobility Network and is currently hired as post-doctoral fellow by SESCAM/Gobierno de Castilla-La Mancha.

References

Arikawa E, Quellhorst G, Han Y, Pan H, Yang J (2011) RT2 profilerTM PCR Arrays: Pathway-focused gene expression profiling with qRT-PCR. *Technical article, SuperArray* Bioscience Corporation.

Ayari B, Soussi-Yanicostas N (2007) FGFR1 and anosmin-1 underlying genetically distinct forms of Kallmann syndrome are co-expressed and interact in olfactory bulbs. Dev Genes *Evol* 217:169–175.

Back SA, Tuohy TM, Chen H, Wallingford N, Craig A, Struve J, Luo NL, Banine F, Liu Y, Chang A, Trapp BD, Bebo BF, Jr., Rao MS, Sherman LS (2005) Hyaluronan accumulates in demyelinated lesions and inhibits oligodendrocyte progenitor maturation. *Nat Med* 11:966-972.

Barkhof F, Bruck W, De Groot CJ, Bergers E, Hulshof S, Geurts J, Polman CH, van d, V (2003) Remyelinated lesions in multiple sclerosis: magnetic resonance image appearance. *Arch Neurol* 60:1073-1081.

Bramow S, Frischer JM, Lassmann H, Koch-Henriksen N, Lucchinetti CF, Sorensen PS, Laursen H (2010) Demyelination versus remyelination in progressive multiple sclerosis. *Brain* 133:2983-2998.

Bribián A, Barallobre MJ, Soussi-Yanicostas N, de Castro F (2006) Anosmin-1 modulates the Fgf-2-dependent migration of oligodendrocyte precursors in the developing optic nerve. *Mol Cell Neurosci* 33:2–14.

Bribián A, Esteban PF, Clemente D, Soussi-Yanicostas N, Thomas JL, Zalc B, de Castro F (2008) A novel role for anosmin-1 in the adhesion. *Dev Neurobiol.* 68:1503-1516.

Cariboni A, Pimpinelli F, Colamarino S, Zaninetti R, Piccolella M, Rumio C, Piva F, et al. 2004. The product of X-linked Kallmann's syndrome gene (KAL1) affects the migratory activity of gonadotropin-releasing hormone (GnRH)-producing neurons. *Hum Mol Genet* 13:2781–2791.

Castronovo V, Luyten F, van den Brule E and Sobel ME (1992). Identification of a 14-kDa laminin binding protein (HLBP14) in human melanoma cells that is identical to the 14-kDa galactoside binding lectin. *Arch. Biochem. Biophys.* 297:132-138.

Chang-Hong R, Wada M, Koyama S, Kimura H, Arawaka S, Kawanami T, Kurita K, Kadoya T, Aoki M, Itoyama Y, Kato T (2005) Neuroprotective effect of oxidized galectin-1 in a transgenic mouse model of amyotrophic lateral sclerosis. *Exp Neurol.* 194:203-211.

Charles P, Reynolds R, Seilhean D, Rougon G, Aigrot MS, Niezgoda A, Zalc B, Lubetzki C (2002) Re-expression of PSA-NCAM by demyelinated axons: an inhibitor of remyelination in multiple sclerosis? Brain 125:1972-1979.

Chung WC, Tsai PS (2010) Role of fibroblast growth factor signaling in gonadotropin-releasing hormone neuronal system development. *Front Horm Res* 39:37-50.

Clemente D, Ortega MC, Arenzana FJ, de CF (2011) FGF-2 and Anosmin-1 are selectively expressed in different types of multiple sclerosis lesions. *J Neurosci* 31:14899-14909.

Colognato H and Yurchenco PD (2000) Form and function: The Laminin family of heterotrimers. *Dev. Dynamics* 218:213-234.

Cooper DNW, Massa SM and Barondes SH (1991). Endogenous muscle lectin inhibits myoblast adhesion. *J. Cell Biol.* 116:1437-1448.

Danguy A, Camby I, Kiss R (2002). Galectins and cancer. *Biochim Biophys Acta* 1572:285-293.

de Castro F, Esteban PF, Bribián A, Murcia-Belmonte V, García-González D, and Clemente D (2013) The adhesion molecule Anosmin-1 in Neurology: Kallmann syndrome and beyond. In (Schousboe A Ed.) *Cell adhesion molecules: implications in neurological diseases.* Springer (New York, U.S.A.)

de Castro F, Bribián A (2005). The molecular orchestra of the migration of oligodendrocyte precursors during development. *Brain Res Brain Res Rev* 49:227–241.

de Morsier G (1954) Studies in cranio-encephalic dysraphia. I. Agenesia of the olfactory lobe (lateral telencephaloschisis) and of the callous and anterior commissures (median telencephaloschisis); olfacto-genital dysplasia. *Schweiz Arch Neurol Psychiatr* 74:309-361.

del Castillo I, Cohen-Salmon M, Blanchard S, Lutfalla G and Petit C (1992) Structure of the X-linked Kallmann syndrome gene and its homologous pseudogene on the Y chromosome. *Nat. Genet.* 2:305–310.

Do KY, Smith DF and Cummings RD (1990) LAMP-1 in CHO cells is a primary carrier of poly-N-acetyllactosamine chains and is bound preferentially by a mammalian S-type lectin. *Biochem. Biophys. Res. Commun.* 173:1123-1128.

Dodé C, Hardelin JP (2009) Kallmann syndrome. *Eur J Hum Genet* 17:139-146.

Dodé C, et al. (2003) Loss-of-function mutations in FGFR1 cause autosomal dominant Kallmann syndrome. *Nat Genet* 33:463-465.

Dubois-Dalcq M, ffrench-Constant C, Franklin RJ (2005) Enhancing central nervous system remyelination in multiple sclerosis. *Neuron* 48:9-12.

Endo Y, Ishiwata-Endo H, Yamada KM (2012) Extracellular matrix protein anosmin promotes neural crest formation and regulates FGF, BMP, and WNT activities. *Dev Cell.* 23:305-316.

Esteban PF, Murcia-Belmonte V, García-González D, de Castro F (2012) *The cysteine Accepted Nov* 2012 by J Neurochem.

Franco B, Guioli S and Pragliola A (1991) A gene deleted in Kallmann's syndrome shares homology with neural cell adhesion and axonal path-finding molecules. *Nature* 353,529-536.

Frohman EM, Racke MK, Raine CS (2006) Multiple sclerosis--the plaque and its pathogenesis. *N Engl J Med* 354:942-955.

Garcia-Alonso L, Fetter RD, Goodman CS (1996) Genetic analysis of Laminin A in Drosophila: extracellular matrix containing laminin A is required for ocellar axon pathfinding. *Development* 122:2611–2621.

Garcıá-González D, Clemente D, Coelho M, Esteban PF, Soussi-Yanicostas N and de Castro F (2010) Dynamic roles of FGF-2 and Anosmin-1 in the migration of neuronal precursors from the subventricular zone during pre- and postnatal development. *Exp. Neurol.* 222:285–295.

Gianola S, de Castro F and Rossi F (2009) Anosmin-1 stimulates outgrowth and branching of developing Purkinje axons. *Neuroscience* 158:570–584.

Gitt MA, Massa SM, Leffler H, Barondes SH (1992) Isolation and expression of a gene encoding L-14-II, a new human soluble lactose-binding lectin. *J. Biol. Chem.* 267:10601-10606.

González-Martínez D, Kim SH, Hu Y, Guimond S, Schofield J, Winyard P, Vannelli GB, et al. (2004) Anosmin-1 modulates fibroblast growth factor receptor 1 in human gonadotropin-releasing hormone olfactory neuroblasts through a heparan sulfate-dependent mechanism. *J Neurosci* 24:10384–10392.

Gu M, Wang W, Song WK, Cooper DN, Kaufman SJ (1994) Selective modulation of the interaction of alpha 7 beta 1 integrin. *J Cell Sci.* 107:175-181.

Han H, Xia Y, Wang S, Zhao B, Sun, Yuan L (2011) Synergistic effects of galectin-1 and reactive astrocytes on functional recovery after contusive spinal cord injury. *Arch Orthop Trauma Surg.* 131:829-839.

Hardelin JP and Dodé C (2008) The complex genetics of Kallmann syndrome: KAL1, FGFR1, FGF8, PROKR2, PROK2, et al. *Sex. Dev.* 2:181–193.

Horie H, Inagaki Y, Sohma Y, Nozawa R, Okawa K, Hasegawa M, Muramatsu N, Kawano H, Horie M, Koyama H, Sakai I, Takeshita K, Kowada Y, Takano M, Kadoya T (1999) Galectin-1 regulates initial axonal growth. *J Neurosci.* 19:9964-9974.

Horie H, Kadoya T, Hikawa N, Sango K, Inoue H, Takeshita K, Asawa R, Hiroi T, Sato M, Yoshioka T, Ishikawa Y. (2004) Oxidized galectin-1 stimulates macrophages. *J Neurosci.* 24:1873-1880.

Horie H, Kadoya T, Sango K, Hasegawa M. (2005) Oxidized galectin-1 is an essential factor for peripheral nerve regeneration. *Curr Drug Targets* 6:385-394.

Hu Y, González-Martínez D, Kim SH, Bouloux PM (2004) Cross-talk of anosmin-1, the protein implicated in X-linked Kallmann's syndrome, with heparan sulphate and urokinase-type plasminogen activator. *Biochem J* 384:495–505.

Hu Y, Guimond SE, Travers P, Cadman S, Hohenester E, Turnbull JE, Kim SH and Bouloux PM (2009) Novel mechanisms of fibroblast growth factor receptor 1 regulation by extracellular matrix protein anosmin-1. *J Biol Chem.* 284:29905–29920.

Hudson ML, Kinnunen T, Cinar HN and Chisholm AD (2006) C. elegans Kallmann syndrome protein KAL-1 interacts with syndecan and glypican to regulate neuronal cell migrations. *Dev. Biol.* 294:352–365.

Ishibashi S, Kuroiwa T, Sakaguchi M, Sun, Kadoya T, Okano H, Mizusawa H (2007) Galectin-1 regulates neurogenesis in the subventricular zone and promotes functional recovery after stroke. *Exp Neurol.* 207:302-313.

Jian B, Nagineni CN, Meleth S, Grizzle W, Bland K, Chaudry I, Raju R (2009) Anosmin-1 involved in neuronal cell migration is hypoxia inducible and cancer regulated. *Cell Cycle* 8:3770-3776.

Kajitani K, Nomaru H, Ifuku M, Yutsudo N, Dan Y, Miura T, Tsuchimoto D, Sakumi K, Kadoya T, Horie H, Poirier F, Noda M, Nakabeppu Y (2009) Galectin-1 promotes basal and kainate-induced proliferation of neural progenitors in the dentate gyrus of adult mouse hippocampus. *Cell Death Differ.* 16:417-427.

Kallmann FJ, Schoenfeld WA, Barrera SE (1944) The genetic aspects of primary eunuchoidism. *Am J Ment Defic* XLVIII:203-236.

Kato T, Kurita K, Seino T, Kadoya T, Horie H, Wada M, Kawanami T, Daimon M, Hirano A (2001) Galectin-1 is a component of neurofilamentous lesions. *Biochem Biophys Res Commun.* 282:166-172.

Kato T, Ren CH, Wada M, Kawanami T. (2005) Galectin-1 as a potential therapeutic agent for amyotrophic lateral sclerosis. *Curr Drug Targets* 6:407-418.

Kuhlmann T, Miron V, Cui Q, Wegner C, Antel J, Bruck W (2008) Differentiation block of oligodendroglial progenitor cells as a cause for remyelination failure in chronic multiple sclerosis. *Brain* 131:1749-1758.

Kurihara D, Ueno M, Tanaka T, Yamashita T (2010) Expression of galectin-1 in immune cells and glial cells after spinal cord injury. *Neurosci Res.* 66:265-270.

Kurushima H, Ohno M, Miura T, Nakamura TY, Horie H, Kadoya T, Ooboshi H, Kitazono T, Ibayashi S, Iida M, Nakabeppu Y (2005) Selective induction of DeltaFosB in the brain after transient forebrain ischemia accompanied by an increased expression of galectin-1, and the implication of DeltaFosB and galectin-1 in neuroprotection and neurogenesis. *Cell Death Differ.* 12:1078-1096.

Lallier TE, Whittaker CA, DeSimone DW (1996) Integrin alpha 6 expression is required for early nervous system development in Xenopus laevis. *Development* 122:2539–2554.

Leffler H and Barondes SH (1986) Specificity of binding of three soluble rat lung lectins to substituted and unsubstituted mammalian β-Galactosides. *J Biol Chem.* 261:10119-10126.

Legouis R, Hardelin JP, Levilliers J et al. (1991) The candidate gene for the X-linked Kallmann syndrome encodes a protein related to adhesion molecules. *Cell* 67:423–435.

Legouis R, Cohen-Salmon M, del Castillo I, Levilliers J, Capy L, Mornow JP, Petit C (1993) Characterization of the chicken and quail homologues of the human gene responsible for the X-linked Kallmann syndrome. *Genomics* 17:516–518.

Liesi P, Kaakkola S, Dahl D & Vaheri A (1984). Laminin is induced in astrocytes of adult brain by injury. *EMBO J.* 3:683-686.

Lopez-Lucendo MF, Solís D, André S, Hirabayashi J, Kasai K, Kaltner H, Gabius HJ and Romero A (2004) Growth-regulatory Human Galectin-1: Crystallographic Characterisation of the Structural Changes Induced by Single-site Mutations and their Impact on the Thermodynamics of Ligand Binding. *J of Mol Biol.* 343:957-970.

Luckenbill-Edds L (1997) Laminin and the mechanism of neuronal outgrowth. *Brain Research Reviews* 23:1-27.

Lutomski D, Joubert-Caron R, Lefebure C, Lefebure C, Salama J, Belin C, Bladier D, Caron M (1997) Anti-galectin-1 autoantibodies in serum of patients with neurological diseases. *Clin Chim Acta* 262:131–138.

Maestre de San Juan A (1856) Teratología: falta total de los nervios olfatorios con anosmia en un individuo en quien existía una atrofia congénita de los testículos y el miembro viril. *El Siglo Médico, Madrid* 3:211-221.

Mahanthappa NK, Cooper DNW, Barondes SH and Schwarting G (1994) Rat olfactory neurons can utilize the endogenous lectin, L-14, in a novel adhesion mechanism. *Development* 120:1373-1384.

McGraw J, Gaudet AD, Oschipok LW, Steeves JD, Poirier F, Tetzlaff W, Ramer MS (2005) Altered primary afferent anatomy and reduced thermal sensitivity in mice lacking galectin-1. *Pain* 114:7-18.

Merkle RK and Cummings RD (1988). Asparagine-linked oligosaccharides containing poly-N-acetyllactosamine chains are preferentially bound by immobilized calf heart agglutinin. *J Biol. Chem.* 263:16143-16149.

Murcia-Belmonte V, Esteban PF, García-González D, De Castro F (2010) Biochemical dissection of Anosmin-1 interaction with FGFR1 and components of the extracellular matrix. *J Neurochem.* 115:1256-1265.

Murtomäki S, Risteli J, Risteli L, Koivisto U-M, Johansson S & Liesi P (1992). Laminin and its neurite outgrowth-promoting domain in the brain in Alzheimer's disease and Down syndrome patients. *J of Neurosci. Res.* 32:261-273.

Offner H, Celnik B, Bringman TS, Casentini-Borocz D, Nedwin GE, Vandenbark AA (1990). Recombinant human beta-galactoside binding lectin suppresses clinical and histological signs of experimental autoimmune encephalomyelitis. *J Neuroimmunol* 28:177–184.

Okubo K, Sakai F, Lau EL, Yoshizaki G, Takeuchi Y, Naruse K, Aida K, Nagahama Y (2006) Forebrain gonadotropin-releasing hormone neuronal development: insights from transgenic medaka and the relevance to X-linked Kallmann syndrome. *Endocrinology* 147:1076-1084.

Pasquini LA, Millet V, Hoyos HC, Giannoni JP, Croci DO, Marder M, Liu FT, Rabinovich GA, Pasquini JM (2011) Galectin-3 drives oligodendrocyte differentiation to control myelin. *Cell Death Differ.* 18:1746-1756.

Patani R, Balaratnam M, Vora A, Reynolds R (2007) Remyelination can be extensive in multiple sclerosis despite a long disease course. *Neuropathol Appl Neurobiol* 33:277-287.

Patnaik SK, Potvin B, Carlsson S, Sturm D, Leffler H, Stanley P (2006) Complex N-glycans are the major ligands for galectin-1, -3, and -8 on Chinese hamster ovary cells. *Glycobiol.* 16:305-317.

Patrikios P, Stadelmann C, Kutzelnigg A, Rauschka H, Schmidbauer M, Laursen H, Sorensen PS, Bruck W, Lucchinetti C, Lassmann H (2006) Remyelination is extensive in a subset of multiple sclerosis patients. *Brain* 129:3165-3172.

Prineas JW, Connell F (1979) Remyelination in multiple sclerosis. *Ann Neurol* 5:22-31.

Puche AC, Poirier F, Hair M, Bartlett PF and Key B (1996) Role of Galectin-1 in the developing mouse olfactory system. *Dev. Biol* 179:274-287.

Qu WS, Wang YH, Ma JF, Tian DS, Zhang Q, Pan DJ, Yu ZY, Xie MJ, Wang JP, Wang W (2011) Galectin-1 attenuates astrogliosis-associated injuries and improves recovery of rats following focal cerebral ischemia. *J Neurochem.* 116:217-226.

Robertson A, MacColl GS, Nash JA, Boehm MK, Perkins SJ, Bouloux PM (2001) Molecular modelling and experimental studies of mutation and cell-adhesion sites in the fibronectin type III and whey acidic protein domains of human anosmin-1. *Biochem J* 357:647–659.

Rorive S, Belot N, Decaestecker C, Lefranc F, Gordower L, Micik S, Maurage CA, Kaltner H, Ruchoux MM, Danguy A, Gabius HJ, Salmon I, Kiss R, Camby I (2001) Galectin-1 is highly expressed in human. *Glia* 33:241-255.

Rugarli EI, Ghezzi C, Valsecchi V and Ballabio A (1996) The Kallmann syndrome gene product expressed in COS cells is cleaved on the cell surface to yield a diffusible component. *Hum. Mol. Genet.* 5:1109–1115.

Sakaguchi M, Shingo T, Shimazaki T, Okano HJ, Shiwa M, Ishibashi S, Oguro H, Ninomiya M, Kadoya T, Horie H, Shibuya A, Mizusawa H, Poirier F, Nakauchi H, Sawamoto K, Okano H (2006). A carbohydrate-binding protein, Galectin-1, promotes proliferation of adult neural stem cells. *Proc Natl Acad Sci USA* 103:7112-7117.

Sakaguchi M, Imaizumi Y, Okano H (2007) Expression and function of galectin-1 in adult neural stem cells. *Cell Mol Life Sci.* 64: 1254-1258.

Sakaguchi M, Imaizumi Y, Shingo T, Tada H, Hayama K, Yamada O, Morishita T, Kadoya T, Uchiyama N, Shimazaki T, Kuno A, Poirier F, Hirabayashi J,Sawamoto K, Okano H (2010) Regulation of adult neural progenitor cells by Galectin-1/beta1 Integrin interaction. *J Neurochem.* 113:1516-1524.

Sakaguchi M, Arruda-Carvalho M, Kang NH, Imaizumi Y, Poirier F, Okano H, Frankland PW (2011) Impaired spatial and contextual memory formation in galectin-1 deficient mice. *Mol Brain* 4:33 (4:1-10).

Santos JF, Müller U (2011) Extracellular matrix functions during neuronal migration and lamination in the mammalian central nervous system. *Dev. Neurobiol* 71:889-900.

Sasaki T, Hirabayashi J, Manya H, Kasai K, Endo T (2004) Galectin-1 induces astrocyte differentiation, which leads to production of brain. *Glycobiology* 14:357-63.

Skrincosky DM, Allen HJ and Bernacki RJ (1993) Galaptin-mediated adhesion of human ovarian carcinoma A121 cells and detection of cellular galaptin-binding glycoproteins. *Cancer Res.* 53:2667-2675.

Soussi-Yanicostas N, de Castro F, Julliard AK, Perfettini I, Chedotal A and Petit C (2002) Anosmin-1, defective in the X-linked form of Kallmann syndrome, promotes axonal branch formation from olfactory bulb output neurons. *Cell* 109:217–228.

Stancic M, van Horssen J, Thijssen VL, Gabius HJ, van der Valk P, Hoekstra D, Baron W. Increased expression of distinct galectins in multiple sclerosis. *Neuropathol Appl* Neurobiol. 37:654-671.

Tengara S, Tominaga M, Kamo A, Taneda K, Negi O, Ogawa H, Takamori K (2010) Keratinocyte-derived anosmin-1, an extracellular glycoprotein encoded by the X-linked Kallmann syndrome gene, is involved in modulation of epidermal nerve density in atopic dermatitis. *J Dermatol Sci* 58:64-71.

Toscano MA, Bianco GA, Ilarregui JM, Croci DO, Correale J, Hernandez JD, Zwirner NW, Poirier F, Riley EM, Baum LG, Rabinovich GA (2007). Differential glycosylation of TH1, TH2 and TH-17 effector cells selectively regulates susceptibility to cell death. *Nat Immunol* 8:825–834.

Van den Brûle FA, Buicu C, Baldet M, Sobel ME, Cooper DN, Marschal P, Castronovo V (1995) Galectin-1 modulates human. *Biochem Biophys Res Commun.* 209:760-767.

Venstrom KA and Reichardt LF (1993) Extracellular matrix. 2: Role of extracellular matrix molecules and their receptors in the nervous system. *FASEB Journal* 7:996-1003.

Wada M, Ono S, Kadoya T, Kawanami T, Kurita K, Kato T (2003) Decreased galectin-1 immunoreactivity *J Neurol Sci.* 208:67-70.

Whitlock KE, Smith KM, Kim H, Harden MV (2005) A role for foxd3 and sox10 in the differentiation of gonadotropin-releasing hormone (GnRH) cells in the zebrafish Danio rerio. *Development* 132:5491-5502.

Yanicostas C, Herbomel E, Dipietromaria A and Soussi-Yanicostas N (2009) Anosmin-1a is required for fasciculation and terminal targeting of olfactory sensory neuron axons in the zebrafish olfactory system. *Mol. Cell. Endocrinol.* 312:53–60.

Yanicostas C, Ernest S, Dayraud C, Petit C and Soussi-Yanicostas N (2008) Essential requirement for zebrafish anosmin-1a in the migration of the posterior lateral line primordium. *Dev. Biol.* 320:469–479.

Zhou Q and Cummings RD (1990). The S-type lectin from calf heart tissue Arch. *Biochem. Biophys.* 281:27-35.

Zhu C, Anderson AC, Schubart A, Xiong H, Imitola J, Khoury SJ, Zheng XX, Strom TB, Kuchroo VK (2005) The Tim-3 ligand galectin-9 negatively regulates T helper type 1 immunity. *Nat Immunol* 6:1245–1252.

In: Laminins: Structure, Biological Activity and Role in Disease ISBN: 978-1-62618-568-5
Editors: Demetrius C. Adams and Emil O. Garcia

Chapter 4

A Potential Role of Laminins and 37 kDa Laminin Receptor Precursor (LRP)/67 kDa Laminin Receptor (LR) in Disease

Jisuk Yun*
Department of Food Safety Evaluation, Korea Food and Drug Administration, Cheongwon-gun, Republic of Korea

Abstract

Laminins, a family of glycoproteins composed of α, ß, and γ chain subunits, present in the basement membrane (BM) of various tissues. Although 16 distinct isoforms have been identified in mammals, their functional differences are not fully understood. Each chain of laminins contains rod-like, globular and coiled regions. Laminin α chain conserves functional sites and chain-specific functional sites.

Laminins play an important role in the organization and structural integrity of BM, and these functions are mediated by various cell surface molecules, such as 37 kDa laminin receptor precursor (LRP)/67 kDa laminin receptor (LR), integrins, dystroglycan, syndecans and lutheran. LR/LRP is a multifunctional protein which acts as a receptor for laminin and also acts as a receptor for many components; carbohydrates, elastin, green tea polyphenol epigallocatechin-3-gallate (EGCG), various viruses and bacteria and prion proteins. LRP/LR can be a factor in pathological processes. In cancer, LRR/LR is upregulated and its overexpression is associated with tumor cell metastasis and attack. EGCG, a component in green tea, represents anti-allergic, anti-tumor and anti-obesity actions via binding to LRP/LR. The functional domain of LRP mediates the anti-cancer action at the physiological concentration of EGCG (0.1-1 μM). In prion disease, LRP/LR acts as a receptor for (abnormal infectious isoform (PrP^{Sc}) as well as a receptor or co-receptor for cellular prion protein (PrP^{C}) and is essential for the propagation and accumulation of PrP^{Sc} in scrapie-infected cells. LRP/LR has also considered as an alternative therapeutic tools in cancer, prion disease and Alzheimer's disease.

* Email: jigel@hanmail.net.

Introduction

Basement membrane (BM) is a dynamic extracellular matrix and contains laminins, type IV collagens, nidogens, heparin sulfate proteoglycans and fibulins. BM provides structural compartmentalization, mechanical support and stability and regulation of cellular activities. BM is connected to cells via integrins and other cell surface receptors, including dystroglycan, syndecans and lutheran, and provides multiple signaling events to cells.

Laminins were originally discovered in a product of mouse Engelbreth-Holm-Swarm sarcoma (EHS), contained large amounts of BM-like materials (Timple et al. 1979). The laminins, a family of multifunctional macromodular glycoproteins, exist in BM and elsewhere. The family of laminin proteins is diverse in structure and biology. Laminins are the most abundant structural non-collagenous glycoproteins in the specialized extracellular moleculs. Laminins contribute to cellular function and tissue morphogenesis, including cell adhesion, migration, growth, differentiation and tumor growth and metastasis both in normal and pathologic processes, either directly by interacting with various cellular surface molecules, or indirectly by trapping growth factors. LRP/LR acts as a receptor for many compounds such as laminins, viruses and prion proteins, and involves in organization of BM. Although many LR segments, such as residues 161-180, also called peptide G, residues 205-229 and C-terminal TEDWS-like repeats, is known as binding sites for laminin, interaction between LPR/LR and laminin has not been fully understood.

This review will provide the current understanding of the possible biosynthesis and structure of laminin and LRP/LR and their potential role in disease, including cancer, prion disease, Alzheimer's disease and other infectious disease.

Laminin Proteins Family and their Function

Laminins are a large family of glycoprotein composed of various combinations of full-length chains in α/ß/γ trimers. In *C. elegans*, laminins have two different isoforms, while in mammals comes in 16 different isoforms from α, ß and γ laminin chains (Table 1). Initially, laminins were named by numbering them in the order of their discovery. To avoid confusion and facilitate transfer of information, a simplified laminin nomenclature assigns a three-digit number, representing the each chain type found within the α/ß/γ trimer (Aumailley et al. 2005). Laminins are a 400-900 kDa in molecular weight and large cross or T-shape molecules with two or three short arms and one long arm (Aumailley *et al.* 2005; Tzu & Peter 2008). The α, ß and γ chains are folded into α helix, accounting for about 30% of conformation (Aumailly & Smyth 1998). The family of laminin proteins is illustrated in Figure 1 and Figure 2. Each laminin chain contains rod-like, globular and coiled regions. Laminin α chains are considered as a functionally important site of the heterotrimers due to their tissue-specific distribution and biological activity. In vascular, the major three type of laminins have been found such as α4 (laminin-411/421), α5 (laminin-511/521) and α3 (laminin-3B11) types. The α5 abundantly exists in the BM of epithelia and blood vessels. The α4 is only distributed in the BM of blood vessels. The α3 is localized in the BM of skin and epithelial, and supports cell adhesion and migration. In the BM and normal tissues, the α3B laminin is co-localized with laminin α4 and α5, but the α3B is shifted to laminin α4 in breast cancer tissue, resulting in interference of tumor angiogenesis (Mori et al. 2011). The shift of vascular laminins is contributed by tumor-associated macrophages; TNF-α significantly suppresses expression of laminin α3B, while interleukin (IL)-1β, tumor necrosis factor (TGF)-α and TGF-β upregulate α4 expression (Mori et al. 2011). Laminin isoforms play unique roles in regulation of the actions of nerve, muscle and Schwann cell: laminin-211(laminin 2), supporting muscle plasma membrane; laminin-221 (laminin 4), regulating adhesion and differentiation of the myotendinous junction; laminin-521 (laminin 11), contributing to nerve terminal differentiation and Schwann cell motility (Patton 2000). In mammalian neuromuscular system, there is expression of seven laminin genes, such as α1, α2, α4, α5, β1, β2 and γ1, producing laminin isoforms, including laminin 111, 211, 221, 411, 421, 511 and 521 (Patton 2000). Laminins are also expressed in Schwann cells of peripheral nervous system (PNS). Laminins and its cellular receptors contribute to Schwann cells development and the suppression of laminins expression damage the proliferation and difference of Schwann cell, followed by apotosis (Yu et al. 2007).

Table 1. Laminin isoforms (Modified from Tzu & Marrinkovich 2008; Hallmann *et al.* 2005; Aumailley et al. 2005)

Name	Composition	Major sites of expression
Laminin-111 (Laminin 1)*	α1 ß1 γ1	Developing epithelial basement menbranes
Laminin-211(Laminin 2)	α2 ß1 γ1	Muscle cells, peripheral nerves
Laminin-121(Laminin 3)	α1 ß2 γ1	Myotendinous junction, placenta
Laminin-221 (Laminin 4)	α2 ß2 γ1	Muscle cells, neuromuscular junction, mesangial matrix of glomertulus
Laminin-332 or 3A32 (Laminin 5or 5A)	α3A ß3 γ2	Epithelial, skin, mammary glands, placenta
Laminin-3B32 (Laminin 5B)	α3B ß3 γ2	Epithelial, skin, uterus, lung
Laminin-311 or 3A11 (Laminin 6or 6A)	α3A ß1 γ1	Epidermis, skin, amnion
Laminin-321 or 3A21 (Laminin 7or 7A)	α3A ß2 γ1	Epidermis, skin, amnion
Laminin-411 (Laminin 8)	α4 ß1 γ1	Endothelial cells, peripheral nerves, smooth muscle, developing kidney and skeletal muscle, neuromuscular junction, mesangial matrix of glomertulus, platelet, white blood cells
Laminin-421(Laminin 9)	α4 ß2 γ1	Vascular endothelial cells, peripheral nerves, muscle fibers, developing kidney and skeletal muscle, platelet, white blood cells
Laminin-511(Laminin 10)	α5 ß1 γ1	Mature endothelial cells, smooth muscle, developing embryo, skin, placenta, blood cells,
Laminin-521(Laminin 11)	α5 ß2 γ1	Mature endothelial cells, smooth muscle, neuromuscular junction, renal glomeruli, placenta, blood cells
Laminin-213(Laminin 12)	α2 ß1 γ3	Kidney capillaries and arterioles, Leydig cells of testis
Laminin-423(Laminin 14)	α4 ß2 γ3	Retina/CNS, hippocampus
Laminin-522 (-)	α5 ß2 γ2	
Laminin-523 (Laminin 15)	α5 ß2 γ3	Retina/CNS

* New nomenclature of laminins (previous name).

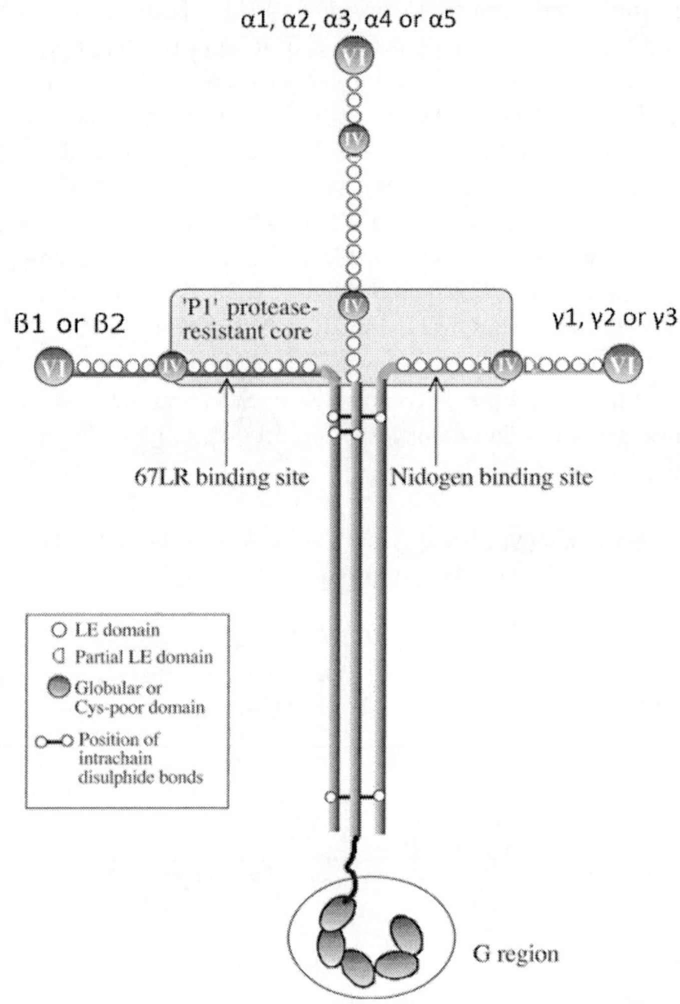

Figure 1. Structure of laminin proteins (modified from Nelson et al. 2008).

Functions of laminins are mediated by various cell surface molecules, such as 37 kDa laminin receptor precursor (LRP)/67 kDa laminin receptor (LR), integrins, dystroglycan, syndecans and lutheran. Globular domains are located in the N- and C-terminal of the laminin α chains, and are essential for interactions with cellular surface receptors (Suzuki et al., 2005). Once these functional sites of laminins bind to the cell surface receptors, intracellular signaling is induced, followed by activating cellular function and tissue morphogenesis, including cell adhesion, migration, growth, differentiation, and tumor growth and metastasis in both normal and pathologic processes (Mori et al. 2011).

Several studies reported that laminin α2, γ1 and γ3 are expressed in many areas of the brain (Colognato & Yurchenco 2000) and laminin α2, α 3 and ß2 are found in tumor microvessels (Simon-Assmann et al. 2011). Jensen et al. (2012) reported that role for laminins as a modulator of proteotoxicity mediated by the neuromuscular junction of BM in *C. elegans* disease models. Laminins also play a role in neurodegeneration diseases related protein misfolding, including Alzheimer's disease and Parkinson's disease (Bonneh-Barkay & Wiley 2009; Jensen et al. 2012).

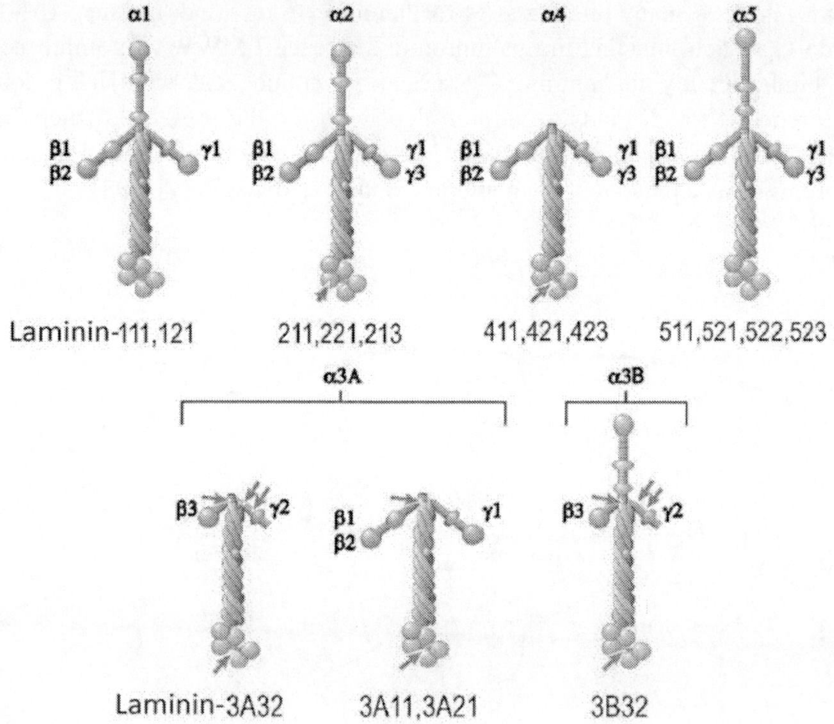

Figure 2. Family of laminin proteins (modified from Aumailley et al. 2005).

Biosynthesis and Structure of 37 kDa Laminin Receptor Precursor (LRP)/67 kDa Laminin Receptor (LR)

LRP forms its mature LR by unknown mechanism involving post-translational fatty acid acylation (Landowski et al. 1995; Nelson et al. 2008; Rieger et al. 1999; Butò et al. 1998) Biosynthesis of LRP/LR is illustrated in Figure 3. Fatty acids, including palmitic acid, oleic acid and stearic acid, acylate LRP and are covalently combined with protein by ester linkage. The acylated LRP can be dimerization either with itself or with another peptide to form its mature LR. The LR associates with laminin-specific integrins in cytoplasm, and is translocated to cell membrane. The C-terminal of LR becomes extracellular by conformational change. Consequently, LR binds to laminin coated on the cell membrane, and laminin increases the translocation of the cytoplasmic receptor complex. LRP also is incorporated into ribosome or translated to nucleus where interacts with histones and DNA.

The amino acid sequence of ribosomal protein SA (*RPSA*), also called LRP/LR gene, shares high homology in mammals (Qiao et al. 2009b). For example, the human *RPSA* sequence shows homology 97.6% with goat and 99% with rat (Zhou et al. 2010; Ménard et al. 1997). The full-length open reading frame (ORF) of human *RPSA* is 888 bp in length, codes a protein of 295 amino acids and is found on chromosome 3p22.2. On the other hand, that of ovine *RPSA* is 885bp in length, encoding 295 amino acids.

As shown Figure 4, many binding sites for laminin are reported: residues 161-180, also called peptide G, which contains the palindromic sequence LMWWML; amino acids 205-229, which binds directly to laminin; C-terminal 53 amino acids containing four acidic TEDWS-like repeats, which binds to laminin depending on the presence of heparin sulfate proteoglycans (HSPGs) (Kazmin et al. 2000; Nelson et al. 1998). Hydrophobic amino acids 86-101 are proposed as a putative transmembrane domain (Ménard et al. 1997).

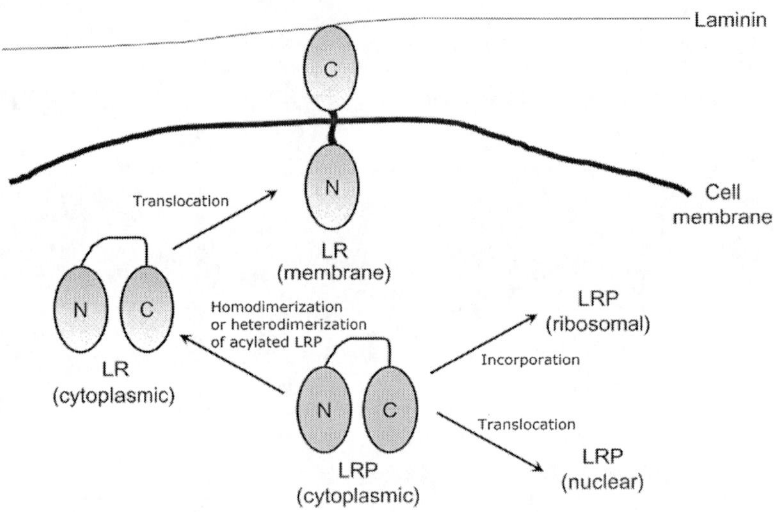

Figure 3. Possible biosynthesis of LRP and LR (modified from Nelson *et al.* 2008).

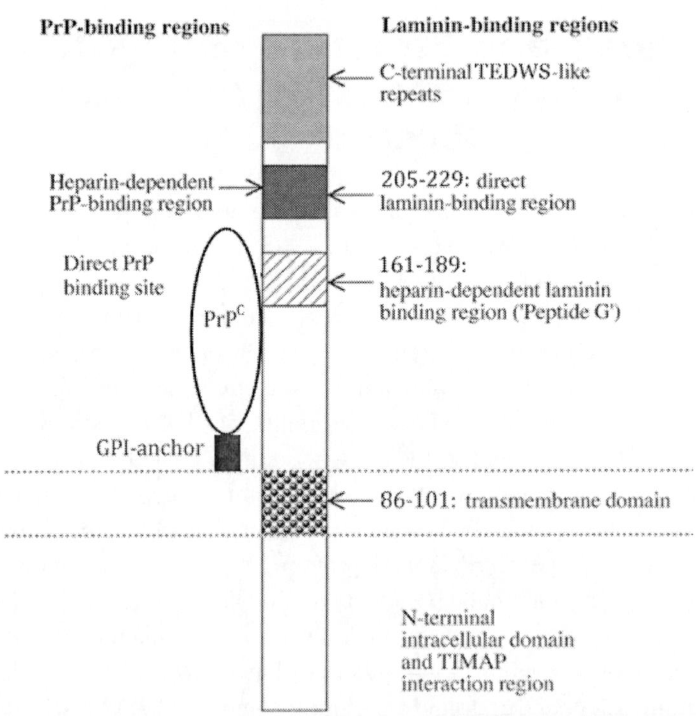

Figure 4. The major function regions and binding sites of LRP/LR in human (Modified from Nelson et al. 2008).

Role of LRP/LR in Disease

LRP/LR, laminin receptor, also acts as a receptor for many components, such as carbohydrates, elastin, viruses, bacteria, prion protein and green tea polyphenol epigallocatechin-3-gallate (EGCG). Laminin-LRP/LR interaction is an important part of mediating changes in the cellular environments, leading to influence on cell adhesion, migration, tumor cell proliferation and metastasis and resistance to apotosis (Formisano et al. 2012). LRP/LR plays a role for factor in pathological processes. EGCG represents anti-allergic, anti-obesity and anti-cancer actions via binding to LRP/LR (Fujimura et al. 2008; Ku et al. 2009; Umeda et al. 2009; Sim & Caughey 2009; Fujimura et al. 2012). Some studies reported potential role of LRP/LR as an alternative therapeutic strategy targeting LRP/LR interactions in cancer, prion disease and Alzheimer's disease (Zuber et al. 2008; Vana et al. 2009).

In Cancer

LRP/LP plays a crucial role in invasiveness, metastasis and growth of tumor. The expression of LR is increased in various cancers, such as carcinoma of in the breast (Martignone et al. 1993), ovary (Song et al. 2012; Liebman et al. 1993; Van den Brûle et al. 1994), uterine (Van den Brûle et al. 1996), prostate (Waltregny et al. 1997), gastric (de Manzoni et al. 1998), colon (Cioce et al.1991; Sanjuan et al. 1996), lung (Satoh et al. 1998), thyroid (Montuori et al. 1999), colon (Cioce et al. 1991), liver (Ozaka et al. 1998) and cervical region (Ménard et al. 1997), while nonintegrin laminin-binding protein such as galectin-3 is decreased (Van den Brûle et al. 1994).

The inverse regulation for the expression of LR and galectin-3 may be due to change in the affinity for laminin-binding. LRR/LR expression is associated with tumor cell metastasis (Nelson et al. 2008; Rieger et al. 1999). In laminin producing tumors, the overexpression of LR is related to a poor prognosis of breast carcinoma (Martignone et al. 1993). The inhibition of LRP/LR-laminin interaction may have a protective effect on tumor growth and metastasis (Kelly et al. 2011).

IPCNNKGAHS, corresponding to peptide LR 161-170 is EGCG binding site, and might be the functional domain responsible for the anti-cancer activity of EGCG (Fujimura et al. 2012).

LR as a cellular surface EGCG receptor inhibits tumor cell proliferation at the physiological concentration of EGCG (0.1-1 μM) (Tachibana et al. 2004). LR mediates anti-inflammatory action of EGCG on lipopolysaccharide (LPS)-stimulated macrophages (Byun et al. 2010) and on LPS-stimulated dendritic cells (Byun et al. 2012).

In Prion Disease

Prion diseases, also called transmissible spongiform encephalopathies (TSEs), are fatal transmissible neurodegenerative disorders with accumulation of an abnormal infectious isoform (PrP^{Sc}) in central nervous system and spleen in humans and animals.

The types of human prion diseases are classified into acquired, iatrogenic and sporadic types according to etiology: acquired, such as kuru, variant Creutzfeldt–Jakob disease (vCJD) and iatrogenic CJD (iCJD); genetic, such as familial CJD (fCJD), Gerstmann-Stra¨ussler-Scheinker syndrome (GSS) and fatal familial insomnia (FFI); sporadic, whose etiology is unknown, which includes sporadic CJD (sCJD) (Prusiner 1991; Aguzzi & Calella 2009). The most common human prion disease is sCJD, accounting for approximately 85%, followed by inherited and acquired by transmission.

The incidence rate of sCJD is about 1-2 per million people a year. The key event in the pathogenesis of prion disease is known as the conformational conversion of naturally expressed cellular prion protein (PrP^{C}) to PrP^{Sc}.

LRP/LR acts as not only as a receptor or co-receptor for PrP^{c} but also a receptor for PrP^{Sc} (Rieger et al. 1997; Rieger et al. 1999; Gauczynski et al. 2001; Gauczynski et al. 2006; Nelson et al. 2008). LRP/LR increases in systems converting PrP^{C} to PrP^{Sc} and in organs accumulating PrP^{Sc} including brain, spleen and pancreas in scrapie-infected rodents including mice and hamsters (Rieger et al. 1997).

As shown Figure 5, PrP^{C} is anchored via a C-terminal glycosylphosphatidyl inositol (GPI) containing one disulfide bridge and five N-terminal octarepeats and co-exists with LRP/LR at the surface of mammalian cells (Gauczynski et al. 2001; Rieger et al. 1999). PrP^{Sc} binds to the cell surface in a specific LRP/LR-heparin sulfate proteoglycan (HSPG)-dependent manner, and then PrP-LRP/LP-HSPG complex is internalized by human enterocytes into endosome/lysosome (Morel et al. 2005; Gauczynski et al. 2006). LRP/LR is essential for the propagation and accumulation of PrP^{Sc} in scrapie-infected cells, and its expression is associated with the degree of PrP^{Sc} propagation (Leucht et al. 2003; Oiao et al. 2009a). Therefore, PrP-LRP/LR interaction is important to pathogenesis of prion disease. Inactivation of LRP/LR interrupts with PrP^{Sc} propagation and PrP^{Sc} accumulation, triggering a prolonged pre-clinical phase and incubation times in scrapie-infected mice (Pflanz et al. 2009a; Pflanz et al. 2009b). Treatment of EGCG, presenting anti-allergic, anti-tumors actions by binding to LRP/LR, impedes PrP^{Sc} formation in scrapie-infected cells and the stress-protective effect of PrP^{c} in uninfected cells (Rambold et al. 2009).

The two PrP-LRP-LR interaction domains on *PRNP* were identified by two-hybrid system in the yeast and cell-binding assay: a direct binding domain amino acids 144-179, also named PrPLRPbd1; an indirect HSPG-dependent LRP-binding domain amino acids 53-93, also called PrPLRPbd2 (Hundt et al. 2001).

As seen in Figure 4, the direct PrP binding site on *RPSA* is amino acids 161-180 and the HSPG-dependent PrP-binding site is amino acids 205-229 (Kazmin et al. 2000). Yun et al. (2011) reported that although the *RPSA* polymorphism at codon 173 does not lead to substitution of amino acid, it remains of interest, because the polymorphism at codon 173 is located in the direct PrP binding domain.

In the regions related to PrP-LRP/LR interaction, ovine *RPSA* polymorphisms did not bring about change in amino acid sequence (Marcos-Carcavilla et al. 2008).

Although on LRP/LR biology or prion biology, the effects of this SNP are not elucidated, it is possible that the differences in amino acids combinations may change the specificity with other proteins, affecting PrP-LRP/LR interaction, strengthening or weakening the species barrier (Marcos-Carcavilla et al. 2008). The *RPSA* amino acid sequences related to resistance barrier to scrapie infection have been identified at positions 241, 271 and 290 in sheep(Zhou et al. 2010), and at positions 241, 272 and 291 in goat (Marcos-Carcavilla et al. 2008).

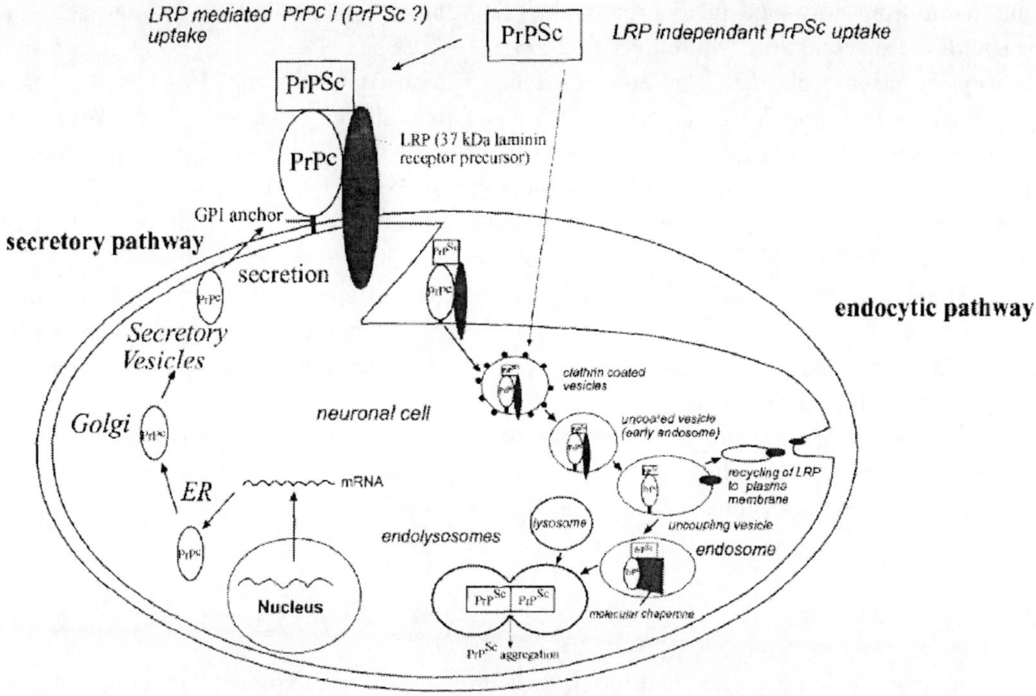

Figure 5. Role of the LRP in the life cycle of prion.

In addition to differences in LRP/LR amino acid sequence, which affects the interaction with PrP^{Sc}, differences in the production and expression of LRP/LR bioavailability on the cell surface and particular conformations might be regional producing cellular differences in binding and accumulation of prion proteins; the result of these differences could lead to various response to scrapie infection (Baloui et al. 2004; Marcos-Carcavilla et al. 2008). LPR is relatively abundant in developing brain tissue, and its expression is limited to a few neurons in cortex known to be especially sensitive to abnormal prion accumulation, and these cells rapidly degenerate during the early stages of CJD (Baloui et al. 2004; Marcos-Carcavilla et al. 2008).

LR is extensively expressed in most adult neurons, and a subset of glial cells and astrocytes, accumulating PrP^{Sc} in scrapie-infected hamsters and mice (Belichenko et al. 1999; Baloui et al. 2004). PrP^{C}-LRP/LR interaction is related to prion protein binding and accumulation, and may occur in the apical dendrites and cell soma (Baloui et al. 2004). LRP presents in the cytoplasm and in the nucleus, while LR exists as a laminin-binding protein on the cell surface or in a free form in the extracellular matrix (Marcos-Carcavilla et al. 2008; Nelson et al. 2008).

As shown Table 2, the genotypic and allelic frequencies of the *RPSA* polymorphisms were not significantly different between controls and sCJD patients. Yun *et al.* (2011) reported that the genotype and allele frequencies of *RPSA* polymorphisms were not influenced by gender.

Some studies suggested that incubation time of prion disease is not affected by sex (Stephenson et al. 2000; Lloyd et al. 2001). On the contrary, the survival time of female mice is shorter than those of male mice (Moreno et al. 2008). These sex effects on prion disease could come from hormonal factors, body size, fat composition, appetite and so on, acting at the end of the survival time (Moreno et al. 2008).

In prion disease, the *PRNP* related resistance and susceptibility may be able to co-exist with other genes modifying its effect (Marcos-Carcavilla et al. 2008). The *PRNP* 219 polymorphism is unique in Asian populations, and *PRNP* 219EK heterozygous genotype has a protective effect on sCJD development (Jeong et al. 2005).

As shown Table 3, there is no significant difference in the genotype and allele frequencies of *RPSA* 5′-UTR -8T>C polymorphism according to the *PRNP* codon 129 or 219 status. These results are might, at least in part, because all sporadic CJD patients had only *PRNP* 129MM or 219EE genotype. Although *RPSA* alone is not a disease-modifying gene for sCJD in Koreans, these SNPs might be linked to the expression or susceptibility of *RPSA* gene and PrP^C-LRP/LR interaction.

PrP^C plays a potential protective role for oxidative stress and has superoxide dismutase (SOD) activity (Sim & Caughey 2009; Bertuchi et al. 2012), while PrP^{Sc} increases intacellular generation of reactive oxygen species (ROS) (Park et al. 2012).

In Alzheimer's Disease

Alzheimer's disease is fatal neurological disorders, forming extracellular senile plaques, such as amyloid β (Aβ) in the afflicted brain. Laminin inhibits formations of Aβ fibrils and depolymerizates Aβ fibrils (Cerpa et al. 2008). LRP/LR and amyloid β precursor protein (APP) are localized in the same cellular surface (Sim & Caughey 2009). Interaction laminin and PrP^C induces neuritogenesis (Graner et al. 2000).

PrP^C overexpression interferes with the β-secrease cleavage of APP, and decreases in Aβ formation in neuronal cells (Parkin et al. 2007).

In Viral and Bacterial Disease

The first step in invasive viral and bacterial pathogens goes into the host cells such as endothelial and epithelial cells. LPR/LR is receptor for various viruses such as Sindbis virus, dengue virus, adeno-associated virus, venezuelean equine encephalitis virus and human immunodeficiency virus (HIV).

Interaction between LR and these viruses or bacteria has not been elucidated.

LR has a high affinity with Sindbis virus, an alphavirus with a wide range of host, and acts as a receptor to enter of the virus into mammalian cells (Wang et al. 1992; Rea et al. 2012).

Overexpression of LR was increased in more infected hamster cells than cells transfected with antisense gene (Wang et al. 1992). LRP/LR acts as a receptor for dengue virus, an arthropod-borne virus transmitted by Aedes mosquitoes (Hidari & Suzuki 2011).

LRP interacts with cytotoxic necrotizing factor 1 (CNF1), a bacterial virulence factor related to the *E. coli* strain, and promotes the internalization of CNF1-positive *E. coli* K1 into the human brain microvascular endothelial cells (HBMEC) (Kim et al. 2005).

Table 2. Genotype and allele frequencies of *RPSA* polymorphisms in controls and sporadic CJD patients (from Yun et al. 2011)

	5′-UTR -8T>C		134-32C>T		519G>A		793+58C>T	
	Controls (n=189)	Sporadic CJDs (n=180)	Controls (n=50)	Sporadic CJDs (n=50)	Controls (n=50)	Sporadic CJDs (n=50)	Controls (n=50)	Sporadic CJDs (n=50)
Genotype								
1/1	91 (48.2%)	87 (48.3%)	33 (66.0%)	33 (66.0%)	34 (68.0%)	33 (66.0%)	28 (56.0%)	24 (48.0%)
1/2	77 (40.7%)	78 (43.3%)	15 (30.0%)	15 (30.0%)	14 (28.0%)	15 (30.0%)	12 (24.0%)	22 (44.0%)
2/2	21 (11.1%)	15 (8.4%)	2 (4.0%)	2 (4.0%)	2 (4.0%)	2 (4.0%)	10 (20.0%)	4 (8.0%)
P-value	0.646		1.000		1.000		0.058	
Allele								
1	279 (0.70)	252 (0.70)	81 (0.81)	81 (0.81)	82 (0.82)	81 (0.81)	68 (0.68)	70 (0.70)
2	119 (0.30)	108 (0.30)	19 (0.19)	19 (0.19)	18 (0.18)	19 (0.19)	32 (0.32)	30 (0.30)
P-value	0.976		1.000		0.856		0.760	

For columns '1' refers to the more common genotype or allele and '2' to the less common genotype or allele of each polymorphism.

Table 3. Genotype and allele frequencies of *RPSA* 5′-UTR -8T>C *1* polymorphism according to the *PRNP* codon 129 or 219 status (from Yun et al. 2011)

PRNP 129	Healthy controls (n=189)			Sporadic CJD patients (n=180)			*P*-value[1]	*P*-value[2]
	MM (n=180)	MV (n=9)	VV (n=0)	MM (n=180)	MV (n=0)	VV (n=0)		
Genotype frequency, n (%)								
TT	84 (46.7)	7 (7.8)	-	87 (48.3)	-	-		
TC	75 (41.7)	2 (2.2)	-	78 (43.3)	-	-	0.233	0.570
CC	21 (11.6)	0 (0)	-	15 (8.3)	-	-		
Allele frequency								
T	0.68	0.89	-	0.70	-	-	0.057	0.469
C	0.32	0.11	-	0.30	-	-		
PRNP 219	Healthy controls (n=189)			Sporadic CJD patients (n=180)			*P*-value[1]	*P*-value[2]
	EE (n=173)	EK (n=16)	KK (n=0)	EE (n=180)	EK (n=0)	KK (n=0)		
Genotype frequency, n (%)								
TT	86 (49.7)	5 (31.3)	-	87 (48.3)	-	-		
TC	68 (39.3)	9 (56.3)	-	78 (43.3)	-	-	0.338	0.600
CC	19 (11.0)	2 (12.4)	-	15 (8.3)	-	-		
Allele frequency								
T	0.69	0.59	-	0.70	-	-	0.244	0.854
C	0.31	0.41	-	0.30	-	-		

[1]Difference in healthy controls with MM and healthy controls with MV.
[2]Difference in healthy controls with MM and sporadic CJD patients with MM.

Conclusion

Laminins are the major glycoproteins of BM and play an important role in cellular function and tissue morphogenesis. LRP/LR, a non-integrin cellular surface receptor, mediates interactions between cells and laminins at a high affinity, and acts as a factor in various pathological processes. Although the function of LRP/LR has not been elucidated, LRP/LR has also considered as an alternative target in therapy of cancer, prion disease and Alzheimer's disease.

Acknowledgment

The author thanks Yong-Sun Kim, MD, Ph. D. and Byung-Hoon Jeong, Ph. D. at Ilsong Institute of Life Science of Hallym University in South Korea.

References

Aguzzi A. & Calella A. M. (2009). Prions: protein aggregation and infectious diseases. *Physiol. Rev.,* 89, 1105-1152.

Alperovitch A., Zerr I., Pocchiari M., Mitrova E., de Pedro Cuesta J., Hegyi I., Collins S., Kretzschmar H., van Duijn C. & Will R. G.(1999). Codon 129 prion protein genotype and sporadic Creutzfeldt-Jakob disease. *Lancet,* 353, 1673-1674.

Aumailly M. & Smyth N. (1998). The role of laminins in basement membrane function. *J. Anat.*, 193, 1-21.

Aumailly M., Bruckner-Tuderman L., Carter W. G., Deutzmann R., Edgar D., Ekblom P., Engel J., Engvall E., Hohenester E., Jones J. C., Kleinman H. K., Marinkovich M. P., Martin G. R., Mayer U., Meneguzzi G., Miner J. H., Miyazaki K., Patarroyo M., Paulsson M., Quaranta V., Sanes J. R., Sasaki T., Sekiguchi K., Sorokin L. M., Talts J. F., Tryggvason K., Uitto J., Virtanen I., von der Mark K., Wewer U. M., Yamada Y. & Yurchenco P. D. (2005). A simplified laminin nomenclature. *Matrix Biol.*, 24, 326-332.

Aumailly M. (2013). The laminin family. *Cell Adh. Migr*. 7, 48-55.

Baloui H., von Boxberg Y., Vinh J, Weiss S., Rossier J., Nothias F. & Stettler O. (2004). Cellular prion protein/laminin receptor: distribution in adult central nervous system and characterization of an isoform associated with a subtype of cortical neurons. *Eur. J. Neurosci.*, 20, 2605-2616.

Bertuchi F. R., Bourgeon D. M., Landemberger M. C., Martins V. R. & Cerchiaro G. (2012). PrPC displays an essential protective role from oxidative stress in an astrocyte cell line derived from PrPC knockout mice. *Biochem. Biophys. Res. Commun.*, 418, 27-32.

Belichenko P. V., Miklossy J., Belser B., Budka H. & Celio M. R. (1999). Early destruction of the extracellular matrix around parvalbumin-immunoreactive interneurons in Creutzfeldt-Jakob disease. *Neurobiol. Dis.*, 6, 269-279.

Bonneh-Barkay D. & Wiley C. A. *(*2009*)*. Brain extracellular matrix in neurodegeneration. *Brain Pathol.*, 19, 573-585.

Butò S., Tagliabue E., Ardini E., Magnifico A., Ghirelli C., van den Brûle F., Castronovo V., Colnaghi M. I., Sobel M. E. & Ménard S. (1998). Formation of the 67-kDa laminin receptor by acylation of the precursor. *J. Cell. Biochem.,* 69, 244-251.

Byun E. H., Fujimura Y., Yamada K. & Tachibana H. (2010). TLR4 signaling inhibitory pathway induced by green tea polyphenol epigallocatechin-3-gallate through 67-kDa laminin receptor. *J. Immunol.*, 185, 33-45.

Byun E. B., Choi H. G., Sung N. Y & Byun E. H. (2012). Green tea polyphenol epigallocatechin-3-gallate inhibits TLR4 signaling through the 67-kDa laminin receptor on lipopolysaccharide-stimulated dendritic cells. *Biochem Biophys. Res. Commun*., 426, 480-485.

Carp R. I., Meeker H., Sersen E. & Kozlowski P. (1998). Analysis of the incubation periods, induction of obesity and histopathological changes in senescence-prone and senescence-resistant mice infected with various scrapie strains. *J. Gen. Virol.,* 79, 2863-2869.

Cerpa W., Dinamarca M. C. & Inestrosa N. C. (2008). Structure-function implication in Alzheimer's disease: effect of Abeta oligomers at central synapses. *Curr. Alzheimer Res*., 5, 233-243.

Cioce V., Castronovo V., Shmookler B. M. Garbisa S., Grigioni W. F., Liotta L. A. & Sobel M. E. (1991). Increased expression of the laminin receptor in human colon cancer. *J. Natl. Cancer Inst.,* 83, 29–36.

Colognato H. & Yurchenco P. D. (2012). Form and function: The laminin family of heterotrimers. *Dev. Dyn*., 218,213-234.

de Manzoni G., Guglielmi A., Verlato G., Tomezzoli A., Pelosi G., Schiavon I. & Cordiano C. (1998). Prognostic significance of 67-kDa laminin receptor expression in advanced gastric cancer. *Oncology*, 55, 456–460.

Formisano P., Ragno P., Pesapane A., Alfano D., Alberobello A. T., Rea V. E., Giusto R., Rossi F. W., Bequinot F., Rossi G. & Montuori N. (2012). PED/PEA-15 interacts with the 67 kDa laminin receptor and regulates cell adhesion, migration, proliferation and apoptosis. *J. Cell. Mol. Med.,* 16, 1435-1446.

Fujimura Y., Umeda D., Yamada K. & Tachibana H. (2008). The impact of the 67 kDa laminin receptor on both cell-surface binding and anti-allergic action of tea catechins. *Arch. Biochem. Biophys*., 476, 133-138.

Fujimura Y., Sumida M., Sugihara K., Tsukamoto S., Yamada K. & Tachibana H. (2012). Green tea polyphenol EGCG sensing motif on the 67-kDa laminin receptor. *PLoS One*, 7, e37942.

Gauczynski S., Peyrin J. M., Haïk S., Leucht C., Hundt C., Rieger R., Krasemann S., Deslys J. P., Dormont D., Lasmézas C. I. & Weiss S. (2001). The 37-kDa/67-kDa laminin receptor acts as the cell-surface receptor for the cellular prion protein. *EMBO J.*, 20, 5863-5875.

Gauczynski S., Nikles D., El-Gogo S., Papy-Garcia D., Rey C., Alban S., Barritault D., Lasmezas C. I. & Weiss S. (2006). The 37-kDa/67-kDa laminin receptor acts as a receptor for infectious prions and is inhibited by polysulfated glycanes. *J. Infect. Dis.,* 194, 702-709.

Graner E., Mercadate A. F., Zanata S. M., Forlenaz O. V., Cabral A. L., Veiga S. S., Juliano M. A., Roesler R., Walz R., Minetti A., Izquierdo I., Martin V. R. & Brentani R. R. (2000). Cellular prion protein binds laminin and mediates neuritogenesis. *Brain Res. Mol. Biochem. Brain Res*., 76, 85-92.

Hallmann, R., Horn, N., Selg, M., Wendler, O., Pausch, F., & Sorokin, L. M. (2005). Expression and function of laminins in the embryonic and mature vasculature. *Physiol. Rev.*, 85, 979-1000.

Hidari K. I. & Suzuki T. (2011). Dengue virus receptor. *Trop. Med. Health*, 39, 37-43.

Hundt C., Peyrin J. M., Haïk S., Gauczynski S., Leucht C., Rieger R., Riley M. L., Deslys J. P., Dormont D., Lasmézas C. I. & Weiss S. (2011). Identification of interaction domains of the prion protein with its 37-kDA/67-kDa laminin receptor. *EMBO J.*, 20, 5876-5886.

Jensen L. T., Møller T. H., Larsen S. M., Jakobsen H. & Olsen A. (2012). A new role for laminins as modulators of protein toxicity in Caenorhabditis elegans. *Aging Cell*, 11, 82–92.

Kazmin D. A., Hoyt T. R., Taubner L., Teintze M. & Starkey J. R. (2000). Phage display mapping for peptide 11 sensitive sequences binding to laminin-1. *J. Mol. Biol.*, 298, 431-445.

Kim K.J., Chung J.W. & Kim K. S. (2005). 67-kDa laminin receptor promotes internalisation of cytotoxic necrotizing factor 1-expressing *E. coli* K1 into human brain microvascular endothelial cells. *J. Biol. Chem.*, 280, 1360-1368.

Ku H. C., Chang H. H., Liu H. C., Hsiao C. H., Lee M. J., Hu Y. J., Hung P. F., Liu C. W. & Kao Y. H. (2009). Green tea (-)-epigallocatechin gallate inhibits insulin stimulation of 3T3-L1 preadipocyte mitogenesis via the 67-kDa laminin receptor pathway. *Am. J. Physiol. Cell. Physiol.*, 297, C121-132.

Jeong B. H., Lee K. H., Kim N. H., Jin J. K., Kim J. I., Carp R. I. & Kim Y. S. (2005). Association of sporadic Creutzfeldt-Jakob disease with homozygous genotypes at PRNP codons 129 and 219 in the Korean population. *Neurogenetics,* 6, 229-232.

Jeong B. H., Nam J. H., Lee Y. J., Lee K. H., Jang M. K., Carp R. I., Lee H. D., Ju Y. R., Ahn J. S., Park K. Y. & Kim Y. S. (2004). Polymorphisms of the prion protein gene (PRNP) in a Korean population. *J. Hum. Genet.*, 49, 319-324.

Kazmin D. A., Hoyt T. R., Taubner L., Teintze M. & Starkey J. R. (2000). Phage display mapping for peptide 11 sensitive sequence binding to laminin-1. *J. Mol. Biol.*, 298, 431-435.

Landowski T. H. & Uthayakumar S. (1995). Studies of the structure of the metastasisi-associated 67 kDa laminin binding protein: fatty acid acylation and evidence supporting dimerization of the 37 kDa gene production to form the mature protein. *Biochem.* 34, 11276-11278.

Leucht C., Simoneau S., Rey C., Vana K., Rieger R., Lasmézas C. I. & Weiss S. (2003). The 37 kDa/67 kDa laminin receptor is required for PrP (Sc) propagation in scrapie-infected neuronal cells. *EMBO Rep.*, 4, 290-295.

Liebman J. M., Burbelo P. D., Yamada Y., Fridman R. & Kleinman H. K. (1993). Altered expression of basement-membrane components and collagenases in ascitic xenografts of OVCAR-3 ovarian cancer cells. *Int. J. Cancer*, 55, 102-109.

Lloyd S. E., Onwuazor O. N., Beck J. A., Mallinson G., Farrall M., Targonski P., Collinge J. & Fisher E. M. (2001). Identification of multiple quantitative trait loci linked to prion disease incubation period in mice. *Proc. Natl. Acad. Sci. USA.*, 98, 6279-6283.

Marcos-Carcavilla A., Calvo J. H., González C., Serrano C., Moazami-Goudarzi K., Laurent P., Bertaud M., Hayes H., Beattie A. E., Lyahyai J., Martín-Burriel I., Torres J. M & Serrano M. (2008). Structural and functional analysis of the ovine laminin receptor gene

(RPSA): Possible involvement of the LRP/LR protein in scrapie response. *Mamm. Genome*, 19, 92-105.

Martignone S., Menard S., Bufalino R., Cascinelli N., Pellegrini R., Taqliabue E., Andreola S. Rilke F. & Colnqqhi M. I. (1993). Prognostic significance of the 67-kilodalton laminin receptor expression in human breast carcinomas. *J. Natl. Cancer Inst.*, 85, 398–402.

Ménard S., Castronovo V., Tagliabue E. & Sobel M. E. (1997). New insights into the metastasis-associated 67 kD laminin receptor. *J. Cell. Biochem*. 67, 155-165.

Montuori N., Muller F., De Riu S., Fenzi G., Sobel M. E., Rossi G. & Vitale M. (1999). Laminin receptors in differentiated thyroid tumors: restricted expression of the 67-kilodalton laminin receptor in follicular carcinoma cells. *J. Clin. Endocrinol. Metab.,* 84, 2086-2092.

Morel E., Andrieu T., Casagrande F., Gauczynski S., Weiss S., Grassi J., Rousset M., Dormont D. & Chambaz J. (2005). Bovine prion is endocytosed by human enterocytes via the 37 kDa/67 kDa laminin receptor. *Am. J. Pathol.*, 167, 1033-1042.

Moreno C. R., Cosseddu G. M., Schibler L., Roig A., Moazami-Goudarzi K., Andreoletti O., Eychenne F., Lajous D., Schelcher F., Cribiu E. P., Laurent P., Vaiman D. & Elsen J. M. (2008). Identification of new quantitative trait Loci (other than the PRNP gene) modulating the scrapie incubation period in sheep. *Genetics*, 179, 723-726.

Mori T., Kariya Y., Komiya E., Higashi S., Miyagi Y., Sekiguci K. & Miyazaki K. (2011). Downregulation of a newly identified laminin, laminin-3B11,in vascular basement membranes of invasive human breast cancers. Cancer Sci., 102, 1095-1100.

Nelson J., McFerran N. V., Pivato G., Chambers E., Doherty C., Steele D. & Timson D. J. (2008). The 67 kDa laminin receptor: structure, function and role in disease. *Biosci. Rep.*, 28, 33-48.

Ozaki I., Yamamoto K., Mizuta T., Kajihara S., Fukushima N., Setoguchi Y., Morito F. & Sakai T. (1998). Differential expression of laminin receptors in human hepatocellular carcinoma. *Gut*, 43, 837-842.

Park Y. G., Jeong J. K., Lee J. H., Lee Y. J., Seol J. W., Kim S. J., Hur T. Y., Jung Y. H., Kang S. J. & Park S. Y. (2012). Lactoferrin protects against prion protein-induced cell death in neuronal cells by preventing mitochondrial dysfunction. *Int. J. Mol. Med.*, 2012 Nov 29. [Epub ahead of print]

Parkin E. T., Watt N. T., Hussain I., Eckman E. A., Eckman C. B., Manson J. C., Baybutt H. N., Turner A. J. & Hooper N. M. (2007). Cellular prion protein regulates beta-secretase cleavage of the Alzheimer's amyloid precursor protein. Proc. Natl. Acad. Sci. USA, 104, 11062-11067.

Patton B. L. (2000). Laminins of the neuromuscular system. *Microsc. Res. Tech.*, 51, 247-261.

Prusiner S. B. (1991). Molecular biology of prion disease. *Science*, 252, 1515-1522.

Pflanz H., Vana K., Mitteregger G., Pace C., Messow D., Sedlaczek C., Nikles D., Kretzschmar H. A. & Weiss S. (2009a). Microinjection of lentiviral vectors expressing small interfering RNAs directed against laminin receptor precursor mRNA prolongs the pre-clinical phase in scrapie-infected mice. *J. Gen. virol.*, 90, 269-274.

Pflanz H., Vana K., Mitteregger G., Renner-Müller I., Pace C., Küchenhoff H., Kretzschmar H. A., Wolf E. & Weiss S. (2009b). Scrapie-infected transgenic mice expressing a laminin receptor decoy mutant reveal a prolonged incubation time associated with low levels of PrPres. *J. Mol. Biol.*, 388:721-729.

Qiao J. W., Su X. O., Li Y. X., Yang J. M., Wang Y. Q., Kouadir M., Zhou X. M., Yang L. F., Yin X. M. & Zhao D. M. (2009a). Variable levels of 37-kDa/67-kDa laminin receptor (RPSA) mRNA in ovine tissues: potential contribution to the regulatory processes of PrP^{Sc} propagation? *Anim. Biotechnol.*, 20, 151-155.

Qiao J., Su X. O., Wang Y., Yang J. M., Kouadir M., Zhou X. M., Yin X. M. & Zhao D. M. (2009b). Cloning and characterization of full-length coding sequence (CDS) of the ovine 37/67-kDa laminin receptor (RPSA). *Mol. Biol. Rep.*, 36, 2131-2137.

Rambold A. S., Miesbauer M., Olschewski D., Seidel R., Riemer C., Smale L., Brumm L., Levy M., Gazit E., Oesterhelt D., Baier M., Becker C. F., Engelhard M., Winklhofer K. F. & Tatzelt J. (2008). Green tea extracts interfere with the stress-protective activity of PrP^C and the formation of PrP^{Sc}. *J. Neurochem.*, 107, 218-229.

Rea V. E., Rossi F. W., de Paulis A., Rango P., Selleri C. & Montuori N. (2012). 67 kDa laminin receptor: structure, function and role in cancer and infection. *Infez. Med.,* 20, 8-12.

Rieger R., Edenhofer F., Lasmézas C. I. & Weiss S. (1997). The human 37-kDa laminin receptor precursor interacts with the prion protein in eukaryotic cells. *Nat. Med.*, 3, 1383-1388.

Rieger R. Lammézas C. I. & Weiss S. (1999). Role of the 37 kDa laminin receptor precursor in the life cycle of prions. *Transfus. Clin. Bil.*, 6, 7-16.

Sanjuan X., Fernandez P. L., Miquel R., Munoz J., Castronovo V., Menard S., Palacin A., Cardesa A. & Campo E. (1996). Overexpression of the 67-kD laminin receptor correlates with tumour progression in human colorectal carcinoma. *J. Pathol.*, 179, 376-380.

Satoh K., Narumi K., Isemura M., Sakai T., Abe T., Matsushima K., Okuda K. & Motomiya M. (1992). Increased expression of the 67kDa-laminin receptor gene in human small cell lung cancer. *Biochem. Biophys. Res. Commun.*, 182, 746-752.

Stephenson D. A., Chiotti K., Ebeling C., Groth D., DeArmond S. J., Prusiner S. B. & Carlson G. A. (2000). Quantitative trait loci affecting prion incubation time in mice. *Genetics*, 69, 47-53.

Sim V. L. & Caughey B. (2009). Recent advances in prion chemotherapeutics. *Infect. Disord. Drug Targets*, 9, 81-91.

Simmon-Assmann P., Orend G., Mammadova-Bach E., Spenle C. & Lefebvre O. (2011). Role of laminins in physiological and pathological angiogenesis. *Int. J. Dev. Biol.*, 55, 455-465.

Song T., Choi C. H., Cho Y. J., Sung C. O., Song S. Y., Kim T. J., Bae D. S., Lee J. W. & Kim B. G. (2012). Expression of 67-kDa laminin receptor was associated with tumor progression and poor prognosis in epithelial ovarian cancer. *Gynecol Oncol.,* 125, 427-32.

Suzuki N., Yokoyama F. & Nomizu M. (2005). Functional sites in the laminin alpha chains. *Connect. Tissue Res.*, 46, 142-152.

Tachibana H., Koga K., Fujimura Y. & Yamada K. (2004). A receptor for green tea polyphenol EGCG. *Mol. Biol.*, 11, 380-381.

Tachibana H. (2009). Molecular Basis for Cancer Chemoprevention by Green Tea Polyphenol EGCG. *Forum Nutr.*, 61, 156–169.

Timpl, R., Rohde, H., Robey, P. G., Rennard, S. I., Foidart, J. M., & Martin, G. R. (1979). Laminin-A glycoprotein from basement membranes. *J. Biol. Chem.*, 254, 9933–9937.

Tzu J. & Marrinkovich M. P. (2008). Bridging structure with function: Structural, regulatory, and developmental role of laminins. *Int. J. Biochem. Cell. Biol.*, 40, 199-214.

Umeda D., Yano S., Yamada K. & Tachibana H. (2008). Green tea polyphenol epigallocatechin-3-gallate signaling pathway through 67-kDa laminin receptor. *J. Biol. Chem.*, 283, 3050-3058.

Vana K., Zuber C., Pflanz H., Kolodziejczak D., Zemora G., Bergmann A. K. & Weiss S. (2009). LRP/LR as an alternative promising target in therapy of prion diseases, Alzheimer's disease and cancer. *Infect. Disord. Drug Targets*, 9, 69-80.

Van den Brûle F. A., Berchuck A., Bast R. C., Liu F. T., Gillet C., Sobel M. E. & Castronovo V. (1994). Differential expression of the 67-kD laminin receptor and 31-kD human laminin-binding protein in human ovarian carcinomas. *Eur. J. Cancer*, 30A, 1096-1099.

Van den Brûle F. A., Buica C., Berchuck A., Bast R. C., Deprez M., Liu F. T., Cooper D. N., Pieters C., Sobel M. E. & Castronovo V. (1996). Expression of the 67-kD laminin receptor, galectin-1, and galectin-3 in advanced human uterine adenocarcinoma. *Hum. Pathol.*, 27, 1185-1191.

Waltregny D,, de Leval L., Menard S,, de Leval J. & Castronovo V. (1997). Independent prognostic value of the 67-kd laminin receptor in human prostate cancer. *J. Natl. Cancer Inst.*, 89, 1224-1227.

Wang K. S., Kuhn R. J., Strauss E. G., Ou S. & Strauss J. H. (1992). High-affinity laminin receptor is a receptor for Sindbis virus in mammalian cells. *J. Virol.*, 66, 4992-5001.

World Health Organization. (cited 2010 Dec 13). Creutzfeldt-Jakob disease(CJD) and variant CJD (vCJF), excerpt from WHO recommended standards and strategies for surveillance, prevention and control of communicable diseases. Available from: http://www. who.int/entity/ zoonoses/diseases/Creutzfeldt.pdf.

Yu W. M., Yu H. & Chen Z. L. (2007). Laminins in peripheral nerve development and muscular dystrophy. *Mol. Neurobiol.*, 35, 288-297.

Yun J., Jeong B. H., Kim H. J., Park Y. J., Lee Y. J., Choi E. K., Carp R., I. & Kim Y. S. (2012). A polymorphism in the *YWHAH* gene encoding 14-3-3 eta that is not associated with sporadic Creutzfeldt-Jakob disease (sCJD). *Mol. Biol. Rep.*, 39, 3619-3625.

Yun J., Jin H. T., Lee Y. J.,, Choi E. K., Carp R., I., Jeong B. H. & Kim Y. S. (2011). The first report of *RPSA* polymorphisms, also called 37/67 kDa LRP/LR gene, in sporadic Creutzfeldt-Jakob disease (CJD). *BMC Med. Gene.*, 12, 108.

Zhou R., Li X., Zheng H., Li L., Quan W., Shen Z., Tong J., Yang Q. & Fu Q. (2010). Molecular Characterization of the Full-Length Coding Sequence of the Caprine Laminin Receptor Gene (RPSA). *Biochem. Genet.*, 48, 962-969.

Zuber C., Mitteregger G., Schuhmann N., Rey C., Knackmuss S., Rupprecht W., Reusch U., Pace C., Little M., Kretzschmar H. A., Hallek M., Büning H. & Weiss S. (2008). Delivery of single-chain antibodies (scFvs) directed against the 37/67 kDa laminin receptor into mice via recombinant adeno-associated viral vectors for prion disease gene therapy. *J. Gen. Virol.*, 89, 2055-61.

In: Laminins: Structure, Biological Activity and Role in Disease ISBN: 978-1-62618-568-5
Editors: Demetrius C. Adams and Emil O. Garcia

Chapter 5

Loss of Laminin Alpha Two

***Mahasweta Girgenrath*[*1] *and Tanya Girgenrath*[2]**
[1]Department of Health Sciences, Boston University, Boston, MA, US
[2]School of Public Health and Professional Degrees,
Tufts University School of Medicine, Boston MA, US

Abstract

Merosin deficient congenital muscular dystrophy type1 (MDC1A) results from mutations in the gene *LAMA2* that encodes for the alpha 2 chain of an indispensible extracellular matrix (ECM) protein known as laminin 211 (also know as merosin). MDC1A is the second most prevalent form of the numerous congenital muscular dystrophies that have been identified. In many cases of this very aggressive disease, premature death of affected children occurs due to either respiratory complications or failure to thrive. While there is no therapy that effectively treats this lethal form of muscular dystrophy, significant progress has been made towards understanding the molecular and biochemical causes for MDC1A and several therapeutic strategies have shown promise in ameliorating the disease pathology in different mouse models of MDC1A. One obvious treatment approach would be to simply replace the missing protein. Indeed, overexpression of ECM proteins such as laminin alpha 1 and agrin significantly improves the pathology of MDC1A since this resolves the primary signaling defect. However, treatment strategies that rely on protein replacement are both challenging to translate into therapy and are often only effective against a singular disease. A more practical and productive approach is to treat some of the secondary manifestations, such as inflammation, necrosis, apoptosis, autophagy, fibrosis and inefficient regeneration of damaged muscle fibers. Not only are these targets more manageable given our current technologies, but they are also common to a host of neuromuscular degenerative diseases, thereby multiplying the potential benefits of an effective treatment. In this review, we will present the diverse strategies that are used to treat the primary defect as well as those that focus on the secondary manifestations of the

[*] Corresponding Author's Contact Information: Mahasweta Girgenrath, Department of Health Science, Boston University, 635 Commonwealth Avenue, Boston, MA 02215, Phone: (617) 353-2739, Fax: (617)-353-7567, E-mail: swetag@bu.edu.

disease. We will also discuss the limits of single-mode therapies that target a single disregulated cellular process such as apoptosis or regeneration. Though they may lead to measurable amelioration, a less than optimal outcome is achieved. MDC1A, like many other dystrophies, has multiple disease drivers. Thus, we can appreciate that a combinatorial strategy that aims to target more than one pathophysiological process may be more beneficial in tackling the complexities of MDC1A pathology. Indeed there is emerging evidence that such strategies result in more comprehensive recovery and therefore could hold promise for MDC1A and other forms of muscular dystrophy.

Congenital Muscular Dystrophy

Congenital Muscular Dystrophies (CMDs) are a highly heterogeneous group of neuromuscular-degenerative disorders that arise from primary gene defects. In general, CMD's are early-onset, autosomal recessive conditions that are characterized by delayed motor development, severe hypotonia, and extreme muscle wasting [1, 2]. Although the first forms of CMD were recognized as early as 1903, it took nearly a century for these conditions to be linked to specific gene defects. There are many different types of CMDs (and the list is still growing) but the most prevalent forms of this disease arise from defects in proteins of the extracellular matrix (ECM) and its associated structures (Fig 1). In skeletal muscle, the interaction between the ECM and the cell membrane (sarcolemma) is of critical importance and any disruption of components of these two structures, or the proteins that allow them to communicate, can lead to muscular dystrophies, a group of inherited disorders that leads to muscle weakness and degeneration.

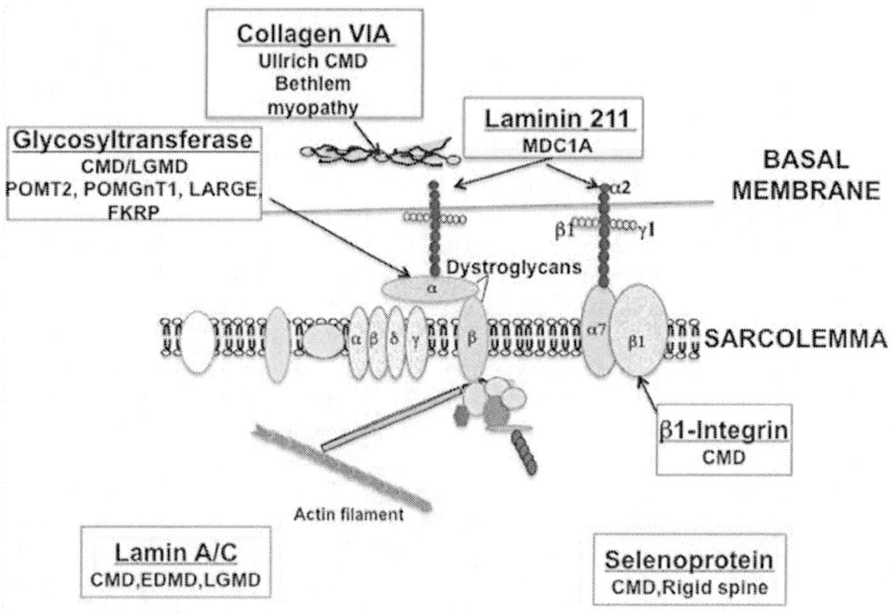

Figure 1. Diagrammatic representation of proteins that are mutated in different forms of congenital muscular dystrophies: collagen V, laminin α-2, integrin α-7, α-dystroglycan, glycosyltransferases POMT1, POMT2, POGnT1, fukutin, FKRP and LARGE, selenoprotein-N and lamin A/C.

Pathology can arise from mutated gene products within the ECM itself, for example defects in the alpha-chain of type VI collagen that cause Ulrich CMD or Bethlem myopathy or disruption of the alpha-two chain of laminin that leads to merosin-deficient congenital muscular dystrophy type 1A (MDC1A). Alternatively, CMDs can arise from mutations in modifying proteins such as glycosyltransferases. These defects lead to abnormalities of O-mannosyl glycosylation of alpha-dystroglycan, an important receptor protein of the dystrophin-glycoprotein complex (DCG) that is instrumental for communication between the sarcolemma and the ECM. In fact, disruption of the DCG complex is also the underlying pathology of the most prevalent and devastating form of muscular dystrophy, Duchenne muscular dystrophy (DMD). Yet another route to CMD involves disruption of proteins associated with the sarcolemma as in the case of an alpha 7integrin mutation that causes a rare form of muscular dystrophy that is restricted to just a few families. Finally, in addition to disruption of cell-to-matrix communication, mutations of certain protein products within the muscle cell can cause CMD. This small subset includes mutations to nuclear proteins as well as proteins associated with the endoplasmic and sarcoplasmic reticulum [2].

The picture that emerges is that CMDs encompass a wide range of pathological conditions that arise from a multitude of different gene and protein defects. This review will focus on the second most prevalent form of CMD, MDC1A, which results from a mutation in laminin. We will explore the molecular defects, clinical features and pathology of MDC1A, as well as discuss the therapeutic approaches that are being investigated to treat this devastating, incurable disease.

Laminin 211

MDC1A is caused by mutations in the LAMA2 gene, which has been mapped to chromosome 6q22-q23 in humans. It encodes for the alpha 2 chain of an essential, multi-functional ECM protein known as laminin 211 [3-5]. The laminin superfamily of matrix proteins play an integral role in multiple cellular processes such as proliferation, differentiation, migration and cell adhesion [6]. Laminins are large, cross-shaped heterotrimeric molecules composed of alpha, beta, and gamma subunits. There are five isoforms of the alpha subunit, and three of the beta and gamma subunits. Each unique laminin is composed of a different combination [7, 8] of these subunits, and to date at least 15 laminin isoforms have been identified that have specific tissue distribution and developmental function [9]. Laminin 211 (composed of $\alpha 2$, $\beta 1$ and $\gamma 1$ subunits) is predominantly expressed in the basal membranes of skeletal muscle, Schwann cells around the peripheral nerves, lamina limitans on the surface of the brain, and capillaries between the astrocyte foot processes and vessels of the brain [2, 3, 10]. This protein is expressed as early as week seven in human embryos and E11 in mice, suggesting that it is important for neuromuscular development [11, 12]. In addition to laminin 211, there are other isoforms of laminin (such as 111, 221, 421 etc) that are co- expressed in skeletal muscle [13, 14]. The laminin protein network in which 211 plays an integral role mediates their effects on muscle cells by interacting with various cell surface receptor proteins.

Binding Partners of Laminin 211

Laminin 211 is associated with a myriad of proteins both in the ECM and in the cell membrane. Specifically, it interacts with the extracellular alpha-subunit of dystroglycan, a transmembrane dimeric protein belonging to the DGC [15]. Alpha-dystroglycan, in turn, connects to the beta-subunit of dystroglycan that is responsible for binding to dystrophin and thus interacts with the intracellular compartment of the myofiber [16]. Yet another important partner of laminin is α7β1integrin, a transmembrane protein complex that links 211 with the underlying cytoskeletal actin network, possibly by interacting with intermediate proteins such as talin and/or integrin-linked kinase [17-20]. Binding of laminin to α-dystroglycan has been shown to activate the PI3K/Akt pathway [21, 22], while interaction of laminin with integrins results in activation of the focal adhesion kinase (FAK) and MAPK pathways that play a role in a multitude of cellular processes.

Interruption of signaling pathways involving α7β1integrin may lead to poor muscle regeneration and cell survival [4, 23, 24]. Likewise, the interaction between laminin and α-dystroglycan is critical for preventing apoptotic muscle cell death [21]. Thus, the integrity of these interfaces is a requisite for physiological muscle function; disruption of signaling mediated through α-dystroglycan or α7β1integrin results in different forms of muscular dystrophy [21, 25-27]. In addition to these cell surface partners, laminin 211 also has many binding partners in the ECM such as agrin, nidogen, parlecan and collagenIV [28, 29].

Pathology of MDC1A

MDC1A is inherited in an autosomal recessive fashion and was first identified by Tome and colleagues in 1994 [30]. Although it is the second most prevalent form of CMD, it was the first to be linked to a genetic defect in *LAMA2* [2]. Since then, many additional mutations have been identified in the *LAMA2* gene that result in a dystrophic phenotype. Mutations that allow for at least partial expression of the alpha-2 chain of laminin display a milder phenotype when compared to severe pathology that results when there is a complete loss of this protein [31, 32]. Based on a population study in Italy, the estimated prevalence and incidence of CMD is 7×10^{-6} and 5×10^{-5}, respectively, almost 40 percent of which is made up of MDC1A patients [33].

Children with MDC1A show profound muscle weakness and hypotonia, either at birth or soon after. They also develop contractures at their elbows, knees and ankles as well as spinal deformities [29]. Serum creatine kinase levels, a standard marker for muscle damage, are also quite high in these patients. Although some affected children may achieve the ability to sit and stand with support, almost none will ever gain independent ambulation. Cranial MRIs from these patients show white matter hypodensity, though impaired brain function is not part of the MDC1A pathology [2, 34].

In most cases of MDC1A, affected children die prematurely due to either respiratory complications or failure to thrive [35, 36]. At present, there is no effective therapy in place to treat this lethal disease.

Mouse Models

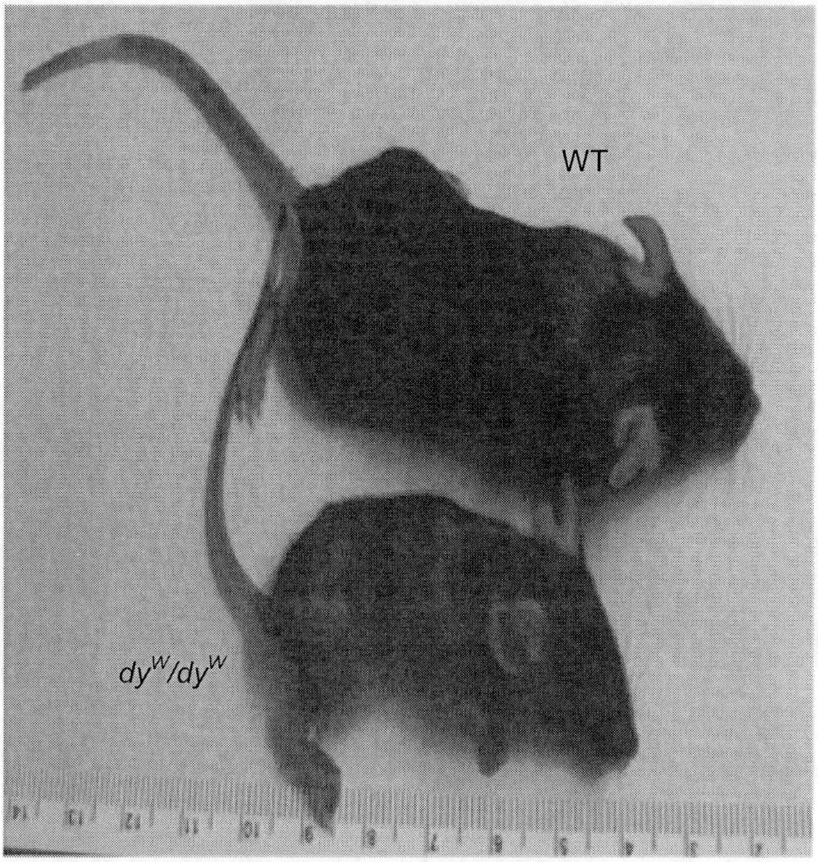

Figure 2. 4-week-old dy^W/dy^W mouse with a healthy littermate. Laminin α-2 deficient mice are much smaller and weaker compared to their healthy littermates. They also develop hind limb paralysis by this age.

There are several mouse models for MDC1A that display a moderate to severe CMD phenotype (directly correlated with the levels of laminin-α2 expressions) [37]. The *dy/dy* and dy^{2J}/dy^{2J} models arose from spontaneous mutations resulting in mice with mild to moderate muscular dystrophy and peripheral neuropathy [38, 39]. While *dy/dy* mouse produce reduced levels of normal protein with unknown mutation, the dy^{2J}/dy^{2J} mouse model contains a spontaneous mutation in the N-terminal domain and expresses reduced levels of laminin-α2 lacking this domain. Another mouse model, dy^{nmj417}/dy^{nmj417}, displays a milder phenotype and has recently been added to the cohort that has a point mutation in the N-terminal domain of laminin-α2 (LN) [40]. Two models with a greatly worsened phenotype have been genetically engineered through homologous recombination [41, 42]. The laminin-α2 null mutant dy^{3K}/dy^{3K} mouse does not express any laminin protein at all and represents the most severe form of MDC1A, while the dy^W/dy^W mouse model expresses a small amount of truncated protein and has been recognized in the recent years as the model that most closely resembles the human pathology (Fig. 2). Like humans, mice that are homozygous for the DyW allele show accelerated muscle degeneration, with limited or no regenerative capacity. They also develop hindlimb paralysis (humans show decreased conduction velocities in the peripheral nerves) as early as four weeks of age and have an extremely short lifespan compared to their healthy littermates [37]. Muscle biopsies of these homozygous mice show a typical dystrophic phenotype, with variable fiber size and vast areas of interstitial spaces filled with infiltrating cells (Fig. 3). In response to ongoing degeneration, biopsies of dystrophic muscles often show an increased number of centrally nucleated fibers (CNFs), a marker of ongoing muscle regeneration. Since laminin deficient muscles have poor regenerative capacities, there are far fewer CNFs in these muscles compared to those seen in muscle biopsies of mdx mice (almost 90% of the fibers are centrally nucleated), a model for DMD.

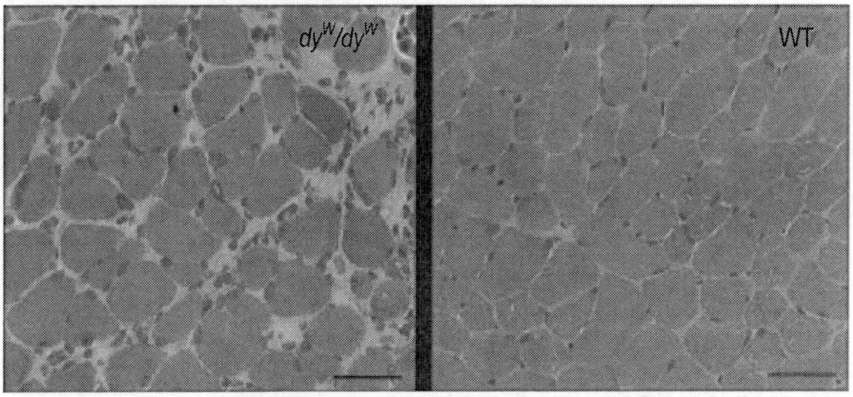

Figure 3. Hematoxylin and eosin staining of tibialis anterior (TA) muscle cross-section of 4-week-old dy^W/dy^W mice show significant pathogenesis with atrophic fibers and large interstitial space filled with infiltrating cells. Bar=50μm.

Pathogenesis of Skeletal Muscle and Potential Therapeutic Targets in MDC1A

Because of its multiple binding partners both in the ECM and the cell membrane, laminin 211 defects result in a major disruption of the structural stability and signal transduction of the ECM [43], the loss of which manifests in failed regeneration, fibrosis, apoptosis, chronic inflammation, and muscle wasting that are typical of MDC1A patients (Fig.4). This creates a devastating domino effect, as all of these secondary pathologies become disease drivers in their own right. One obvious treatment approach would be to simply replace the defective laminin 211 proteins. Indeed, gene or protein replacement to correct the primary cause of MDC1A shows great promises in pre-clinical studies; however, these therapies have translational challenges and are still several years away [23, 44-49]. Currently, a more realistic approach is to treat some of the secondary manifestations of laminin-α2 deficiency, such as apoptosis, inefficient regeneration, inflammation and fibrosis of damaged muscle tissue [44]. Several studies targeting these pathologies have lead to measurable improvement in muscle function, albeit none of them cured it fully. In the next few paragraphs we will discuss some of the therapeutic strategies that are designed to correct primary defects as well as those focused on attenuation of the secondary manifestations.

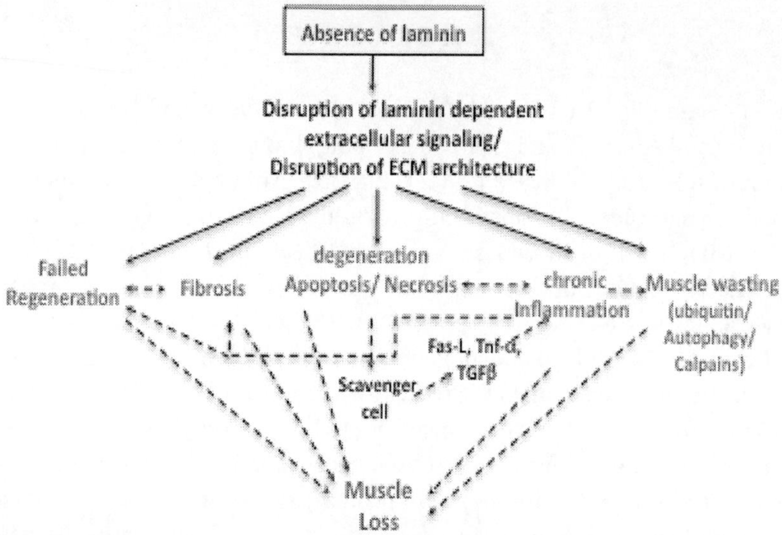

Figure 4. Schematic representation of linking loss of laminin to an array of disregulated cellular processes. Once disregulated, these pathological processes can further amplify each other directly or indirectly and facilitate muscle loss.

Gene and Protein Replacement Therapy

Gene or protein replacement to correct the primary cause of MDC1A shows promise in pre-clinical studies. Overexpression of the laminin $\alpha 1$, $\alpha 2$ chains as well as mini-agrin, a protein that is primarily found in neuromuscular junction, has been shown to partially compensate for a laminin-$\alpha 2$ deficiency. A likely mechanism for this is that they have a stabilizing impact on the signaling pathways involved the dystrophin-dystroglycan complex and α7β1 integrin [23, 45, 47, 49, 50]. In a proof of concept study, Moll et al. elegantly demonstrated that muscle-specific overexpression of mini-agrin, which binds to only one partner of laminin-$\alpha 2$, dystroglycan, is sufficient to partially attenuate the muscle pathology and increase life span in dy^{W}/dy^{W} mice [51, 52]. Later, the same group showed that this therapeutic strategy was effective in slowing the progression of the disease – even when the treatment began after the onset of the disease symptoms [52]. Similarly, transgenic expression of laminin-$\alpha 1$ (normally expressed during embryonic development but not in adult skeletal muscle) in muscle as well as peripheral nerves of laminin-deficient mice has shown marked improvement in muscle and nerve pathology, respectively [48-50], likely by restoring integrin-mediated signaling [53]. Subsequently, in another proof of concept study, Rooney et al. have demonstrated the translational potential of this therapy by systemically delivering Engelbreth-Holm-Swarm-derived mouse laminin-111 protein to dy^{W}/dy^{W} mice. Mice treated with laminin-111 showed marked improvement in survivability and muscle pathology [54]. While restoring the signaling cascade is promising and may be able to correct the aforementioned cellular processes, approaches that rely on protein replacement are challenging to translate into therapy. Another approach, then, is to treat some of the secondary manifestations of MDC1A, such as apoptosis, inefficient regeneration, inflammation and fibrosis of damaged muscle tissue [44].

Apoptosis

Laminin 211 signaling, mediated by both α-dystroglycan and α7β5 integrin, has been shown to promote cell survival, so it is not surprising that apoptotic nuclei are evidenced both in human and mouse muscle biopsies [19, 21, 24, 55]. Although it is possible that, similar to dystrophinopathy, sarcolemmal instability could lead to contraction-related muscle degeneration in MDC1A [56], it was suggested by Vachon *et al.* [19] that laminin deficient cells undergo detachment induced apoptosis due to lack of integrin signaling. Indeed, subsequent work using a zebrafish model of MDC1A suggests that detachment of sarcolemma from the ECM leads to degeneration of muscle fibers (likely due to loss of integrin signaling) independent of the sarcolemmal integrity [57]. The notion that apoptosis plays a significant role in the progression of MDC1A pathology was confirmed by studies that showed both muscle specific overexpression of human Bcl2 (an antiapoptotic protein) and global inhibition of Bax (a pro-apoptotic protein) greatly improved the lifespan, growth, and skeletal muscle histopathology in dy^W/dy^W mice [58, 59]. Neuropathy, however, was only improved in response to global inhibition of Bax. It is interesting that crossing these Bcl2 overexpressing mice with mdx mice, a model of dystrophinopathy, had absolutely no benefit. This result suggests that apoptosis is likely not an important component of this disease [59]. In an effort to identify a pharmaceutical approach to prevent apoptosis, dy^W/dy^W mice were treated with doxycycline, an antibiotic drug that is also known for antiapoptotic effects. Doxycycline treated animals showed a significant increase in lifespan and improved muscle pathology with reduced inflammation [60]. Subsequently, inhibition of the pro-apoptotic glyceraldehype-3-phosphate dehydrogenase (GAPDH)-Siah1-CBP/p300-p53 pathway with Omigapil (antiapoptotic drug being considered for CMD clinical trial) has also shown improved outcomes of the dy^W/dy^W mice [61]. Although anti-apoptotic therapy shows promise, it is important to point out that mice receiving genetic treatment (with Bax inhibition) or pharmaceutical treatment (with doxycycline) did not show any improvement in the regenerative index and displayed persistent inflammation and fibrosis, suggesting that inhibition of apoptosis alone is not sufficient to address the ongoing muscle pathology.

Regeneration

The inability of skeletal muscle to regenerate successfully in response to injury greatly worsens the dystrophic pathology in laminin-α2 deficient mice and likely plays a significant role in disease progression in MDC1A patients. Several growth factors including insulin-like growth factor (IGF-1), basic fibroblast growth factor (bFGF), and nerve growth factor (NGF) have been shown to enhance muscle regeneration in mice [62-66]. IGF-1 signaling plays an important role in the maintenance of muscle mass and is known to improve processes such as proliferation, differentiation, cell survival, and protein synthesis [67]. IGF-1 has been shown to mediate hypertrophy and cell survival by activation of phosphatidyl-inositol 3-kinase (PI3-kinase) and its downstream targets [68, 69], as well as facilitate proliferation via RAF-1/MEK/ERK pathway activation. Muscle-specific overexpression of IGF-1 or administration of high and constant levels of IGF-1 has been shown to cause muscle hypertrophy, increased force production, and improved neurological function in mouse models for DMD and spinal and bulbar muscle atrophy [68, 70]. A short-term study in *dy/dy* model of laminin deficiency previously demonstrated that systemic administration of IGF-1 also improves muscle size and force production [71]. Interestingly, clenbuterol, a beta-agonist acting through β2-adrenoreceptor, also significantly improves muscle mass and force in dy/dy mice [72]. These effects of catecholamines such as clenbutrol could be mediated by activation of IGF-1 signaling, since hypertrophy mediated by clenbuterol is dampened by rapamycin, an inhibitor of Akt/mTOR pathway [67, 73-75]. We have recently reported that muscle-specific overexpression of IGF-1 in dy^W/dy^W mice improved lifespan, increased body mass and muscle mass as well as improved muscle function. These marked improvements could be explained by increased regenerative capacities, as evidenced by an increased number of CNFs and proliferation of myogenic cells [76]. Increlex is a recombinant human IGF-1 that is FDA-approved and presently being tested in a clinical trial in DMD. It is likely that this drug could also have translational potential in the context of MDC1A, with one caveat that unlike the mouse model of DMD, the IGF-1 transgene did not reduce fibrosis and inflammation in dy^W/dy^W muscles.

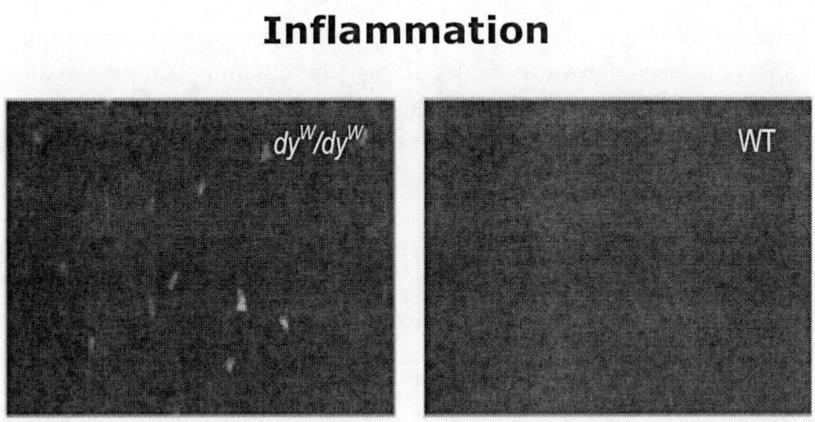

Figure 5. A crossectional view of tibialis anterior muscle stained with Mac-1, a marker for macrophages, (green) shows increased number of inflammatory cells in 3-week-old dy^W/dy^W mice compared to its wild-type littermate. Nuclei are stained with DAPI (blue). Images were captured at 20X magnification.

In MDC1A patients, there is a massive acute inflammatory episode between 2 and 6 months after birth [77]. Chronic signs of inflammation are also evident in muscles of laminin-α2 deficient mice (Fig. 5) [60, 61, 78]. Preliminary data from our laboratory shows early onset of inflammation, with increase in several cytokines including TNF-α, IFNγ, and interleukins in the muscles as well as in the serum of DyW mice. These cytokines have been implicated in mediating inflammation, often through activation of NFκB family members [79]. The NFκB pathway has been found to be strongly upregulated in different muscle wasting conditions including dystrophies [80-84]. We have observed significant increases in NFκB DNA binding and target gene expression in muscles of DyW mice (preliminary results). Indeed, work from our own laboratory and that of Dr. Dennis Guttridge have shown amelioration of pathology in dy^W/dy^W mice treated with a pharmacological inhibitor of NFκB, Nemo Binding Domain (NBD) protein [78]. NBD has previously been shown to be efficacious in treating mouse models of DMD [85]. Interestingly, in addition to inhibition of apoptosis, doxycycline treatment markedly reduced inflammation in dy^W/dy^W mice [60].

Fibrosis

Chronic inflammation and extensive accumulation of fibrotic tissue in the interstitial space are pathological signatures of laminin-α2-deficient muscles [86] (Fig.6). In contrast to DMD, where fibrosis develops later during the disease progression, extensive fibrosis is a key feature of very early MDC1A pathology [86-88] and possibly plays a more etiological role in this disease compared to classical dystrophies, such as Duchenne. Taniguchi *et al.* [86] showed that expression of several ECM components were consistently high in MDC1A patient muscle biopsies. TGF-β has been found to act as a pro-fibrotic cytokine that acts by by increasing the production of matrix proteins in multiple tissues, including the lungs and kidneys [89-91]. In dy^{2J}/dy^{2J} mice, treatment with halofuginone, a compound that reduces collagen synthesis, leads to improvements in muscle cell proliferation and reduced collagen deposition in interstitial space [92]. Additionally, administration of losartan (an angiotensin receptor inhibitor) has been shown to reduce fibrosis by blunting TGF-β signaling in mouse models of Marfan syndrome and DMD [93], but it also leads to marked reduction of fibrosis in both dy^{2J}/dy^{2J} and dy^W/dy^W mouse models [88, 94, 95]. It should be noted, however, that antifibrotic treatment did not result in improved weight gain in any of these studies. Since maintenance of critical body weight is a major challenge for MDC1A patients, a treatment strategy that targets only fibrosis may not be useful as a standalone therapy but could largely help if combined with another therapy. An ideal treatment strategy may be one that targets fibrosis alongside improving muscle mass. This could be achieved by administering losartan in combination with IGF-1 (remember that treatment with IGF alone improves muscle mass but does not address fibrosis).

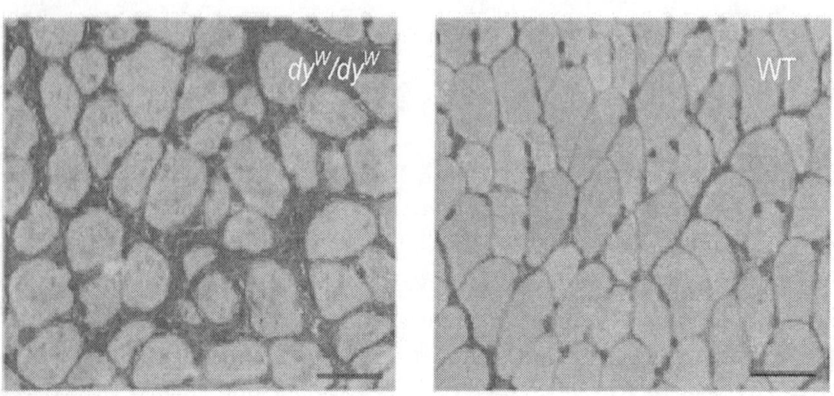

Figure6. Picrosirius Red staining (stain's collagen fibers) of tibialis anterior muscle of 3-week-old dy^W/dy^W mice show extensive fibrosis compared to its wild-type littermate. Bar=50μm.

Two Pathways of Protein Degradation

Cells degrade and recycle proteins and organelles by utilizing the ubiquitin-proteasome system (UPS) and autophagy–lysosomal pathways. MAFbx/atrogin-1 and muscle ring finger protein 1 (MuRF1) are two ubiquitin ligases that tag several molecules of ubiquitin to proteins, which once targeted are degraded by UPS. Levels of both these proteins are largely increased in several scenarios of muscle wasting, including in laminin deficient dystrophies [76, 96-98]. Carmignac *et al.* showed that use of the proteasome inhibitor MG-132 improved lifespan and muscle phenotype in the dy^{3K}/dy^{3K} mice [98].

Autophagocytosis is a catabolic process that utilizes the lysosomal machinery to degrade aged proteins and cellular components by engulfing them in a specialized double membrane structure called autophagosome [99]. Although autophagy is important for the basal homeostasis of cells, dysregulation of this system has been associated with a number of diseases.

Again, Carmignac et al. demonstrated that several key genes integral to autophagic process are upregulated in dy^{3K}/dy^{3K} mice and inhibition of autophagy improves the muscle phenotype in these animals [100].

Combinatorial Therapy

The many pathological symptoms of MDC1A make it a disease with multiple potential therapeutic targets. Using gene and protein replacement therapies to correct the primary cause of MDC1A, the mutation in the LAMA2 gene, would be ideal; however, there are still some years of research required to bring this technology to the clinic.

Meanwhile, a logical alternative is to treat the most prominent symptom of the disease: muscle wasting. As we have described, targeting major drivers of MDC1A pathology such as limited regenerative capacity, uncontrolled apoptosis, fibrosis, and inflammation does indeed lead to measurable improvement in the mouse models of MDC1A, with a caveat that each of these individual treatments results only in partial amelioration of the pathology.

Thus, an emerging idea in the field of MDC1A and other dystrophies is that of a combinatorial therapeutic strategy. A treatment approach that targets more than one disease driver has the potential to achieve a level of pathology-amelioration beyond what can be attained when targeting one driver alone. In fact, Meinen et al. [101] recently showed that combining antiapoptotic therapy with expression of miniagrin results in an additive effect on regenerative and force producing capacities in dy^W/dy^W mice. Preliminary results from our own laboratory show that IGF-1 overexpression in the context of either inhibition of apoptosis or inflammation leads to a marked improvement in the muscle phenotype and locomotory functions of dy^W/dy^W mice.

Thus, simultaneously targeting more than one of the dysregulated cellular mechanisms contributing to the complex pathology seen in MDC1A may be necessary for an effective treatment that addresses the disease symptoms.

Conclusion

The realm of congenital muscular dystrophies is nothing short of vast, and our current capacity to treat these devastating conditions remains limited. However, understanding the underlying mechanisms of these pathologies as laid out in this chapter is necessarily the first step towards finding a cure. With the current technology available to us, an appreciation for the signaling pathways that are dysregulated as a result of the loss or mutation of individual genes allows us to target and remedy the downstream symptoms of CMDs like MDC1A. Using combinatorial approaches, we can significantly improve the standard of living for these patients until the technique of protein and gene replacement therapy is perfected.

Acknowledgment

Photographs of muscle biopsies were taken by Jenny Yamauchi.

References

[1] Schessl, J., Y. Zou, and C.G. Bonnemann, Congenital muscular dystrophies and the extracellular matrix. *Semin Pediatr Neurol*, 2006. 13(2): p. 80-9.

[2] Collins, J. and C.G. Bonnemann, Congenital muscular dystrophies: toward molecular therapeutic interventions. *Curr Neurol Neurosci Rep*, 2010. 10(2): p. 83-91.

[3] Colognato, H. and P.D. Yurchenco, Form and function: the laminin family of heterotrimers. *Dev Dyn,* 2000. 218(2): p. 213-34.

[4] Burkin, D.J. and S.J. Kaufman, The alpha7beta1 integrin in muscle development and disease. *Cell Tissue Res*, 1999. 296(1): p. 183-90.

[5] Leivo, I. and E. Engvall, Merosin, a protein specific for basement membranes of Schwann cells, striated muscle, and trophoblast, is expressed late in nerve and muscle development. *Proc Natl Acad Sci U S A,* 1988. 85(5): p. 1544-8.

[6] Gullberg, D., C.F. Tiger, and T. Velling, Laminins during muscle development and in muscular dystrophies. *Cell Mol Life Sci,* 1999. 56(5-6): p. 442-60.

[7] Aumailley, M., et al., A simplified laminin nomenclature. *Matrix Biol*, 2005. 24(5): p. 326-32.

[8] Miner, J.H. and P.D. Yurchenco, Laminin functions in tissue morphogenesis. *Annu Rev Cell Dev Biol,* 2004. 20: p. 255-84.

[9] Ekblom, P., P. Lonai, and J.F. Talts, Expression and biological role of laminin-1. *Matrix Biol*, 2003. 22(1): p. 35-47.

[10] Feltri, M.L. and L. Wrabetz, Laminins and their receptors in Schwann cells and hereditary neuropathies. *J Peripher Nerv Syst,* 2005. 10(2): p. 128-43.

[11] Patton, B.L., et al., Distribution and function of laminins in the neuromuscular system of developing, adult, and mutant mice. *J Cell Biol*, 1997. 139(6): p. 1507-21.

[12] Sewry, C.A., M. Chevallay, and F.M. Tome, Expression of laminin subunits in human fetal skeletal muscle. *Histochem J*, 1995. 27(7): p. 497-504.

[13] Petajaniemi, N., et al., Localization of laminin alpha4-chain in developing and adult human tissues. *J Histochem Cytochem*, 2002. 50(8): p. 1113-30.

[14] Tiger, C.F., et al., Presence of laminin alpha5 chain and lack of laminin alpha1 chain during human muscle development and in muscular dystrophies. *J Biol Chem*, 1997. 272(45): p. 28590-5.

[15] Ervasti, J.M. and K.P. Campbell, A role for the dystrophin-glycoprotein complex as a transmembrane linker between laminin and actin. *J Cell Biol,* 1993. 122(4): p. 809-23.

[16] Campbell, K.P. and S.D. Kahl, Association of dystrophin and an integral membrane glycoprotein. *Nature*, 1989. 338(6212): p. 259-62.

[17] Wang, H.V., et al., Integrin-linked kinase stabilizes myotendinous junctions and protects muscle from stress-induced damage. *J Cell Biol,* 2008. 180(5): p. 1037-49.

[18] Song, W.K., et al., H36-alpha 7 is a novel integrin alpha chain that is developmentally regulated during skeletal myogenesis. *J Cell Biol,* 1992. 117(3): p. 643-57.

[19] Vachon, P.H., et al., Integrins (alpha7beta1) in muscle function and survival. Disrupted expression in merosin-deficient congenital muscular dystrophy. *J Clin Invest,* 1997. 100(7): p. 1870-81.

[20] Conti, F.J., et al., Talin 1 and 2 are required for myoblast fusion, sarcomere assembly and the maintenance of myotendinous junctions. *Development*, 2009. 136(21): p. 3597-606.

[21] Langenbach, K.J. and T.A. Rando, Inhibition of dystroglycan binding to laminin disrupts the PI3K/AKT pathway and survival signaling in muscle cells. *Muscle Nerve,* 2002. 26(5): p. 644-53.

[22] Xiong, Y., Y. Zhou, and H.W. Jarrett, Dystrophin glycoprotein complex-associated Gbetagamma subunits activate phosphatidylinositol-3-kinase/Akt signaling in skeletal muscle in a laminin-dependent manner. *J Cell Physiol*, 2009. 219(2): p. 402-14.

[23] Rooney, J.E., et al., Laminin-111 restores regenerative capacity in a mouse model for alpha7 integrin congenital myopathy. *Am J Pathol,* 2009. 174(1): p. 256-64.

[24] Laprise, P., et al., Merosin-integrin promotion of skeletal myofiber cell survival: Differentiation state-distinct involvement of p60Fyn tyrosine kinase and p38alpha stress-activated MAP kinase. *J Cell Physiol,* 2002. 191(1): p. 69-81.

[25] Muntoni, F., et al., Defective glycosylation in congenital muscular dystrophies. *Curr Opin Neurol*, 2004. 17(2): p. 205-9.

[26] Cote, P.D., et al., Chimaeric mice deficient in dystroglycans develop muscular dystrophy and have disrupted myoneural synapses. *Nat Genet,* 1999. 23(3): p. 338-42.

[27] Barresi, R. and K.P. Campbell, Dystroglycan: from biosynthesis to pathogenesis of human disease. *J Cell Sci,* 2006. 119(Pt 2): p. 199-207.

[28] Hamill, K.J., et al., Laminin deposition in the extracellular matrix: a complex picture emerges. *J Cell Sci,* 2009. 122(Pt 24): p. 4409-17.

[29] Gawlik, K.I. and M. Durbeej, Skeletal muscle laminin and MDC1A: pathogenesis and treatment strategies. *Skelet Muscle*, 2011. 1(1): p. 9.

[30] Tome, F.M., et al., Congenital muscular dystrophy with merosin deficiency. *C R Acad Sci III,* 1994. 317(4): p. 351-7.

[31] Allamand, V. and P. Guicheney, Merosin-deficient congenital muscular dystrophy, autosomal recessive (MDC1A, MIM#156225, LAMA2 gene coding for alpha2 chain of laminin). *Eur J Hum Genet*, 2002. 10(2): p. 91-4.

[32] Geranmayeh, F., et al., Genotype-phenotype correlation in a large population of muscular dystrophy patients with LAMA2 mutations. *Neuromuscul Disord,* 2010. 20(4): p. 241-50.
[33] Mostacciuolo, M.L., et al., Genetic epidemiology of congenital muscular dystrophy in a sample from north-east Italy. *Hum Genet,* 1996. 97(3): p. 277-9.
[34] Reed, U.C., et al., Congenital muscular dystrophy with cerebral white matter hypodensity. Correlation of clinical features and merosin deficiency. *Brain Dev*, 1996. 18(1): p. 53-8.
[35] Philpot, J., et al., Feeding problems in merosin deficient congenital muscular dystrophy. *Arch Dis Child,* 1999. 80(6): p. 542-7.
[36] Muntoni, F. and T. Voit, The congenital muscular dystrophies in 2004: a century of exciting progress. *Neuromuscul Disord,* 2004. 14(10): p. 635-49.
[37] Guo, L.T., et al., Laminin alpha2 deficiency and muscular dystrophy; genotype-phenotype correlation in mutant mice. *Neuromuscul Disord,* 2003. 13(3): p. 207-15.
[38] Sunada, Y., et al., Deficiency of merosin in dystrophic dy mice and genetic linkage of laminin M chain gene to dy locus. *J Biol Chem,* 1994. 269(19): p. 13729-32.
[39] Sunada, Y., et al., Identification of a novel mutant transcript of laminin alpha 2 chain gene responsible for muscular dystrophy and dysmyelination in dy2J mice. *Hum Mol Genet,* 1995. 4(6): p. 1055-61.
[40] Patton, B.L., et al., A single point mutation in the LN domain of LAMA2 causes muscular dystrophy and peripheral amyelination. *J Cell Sci,* 2008. 121(Pt 10): p. 1593-604.
[41] Kuang, W., et al., Merosin-deficient congenital muscular dystrophy. Partial genetic correction in two mouse models. *J Clin Invest,* 1998. 102(4): p. 844-52.
[42] Miyagoe, Y., et al., Laminin alpha2 chain-null mutant mice by targeted disruption of the Lama2 gene: a new model of merosin (laminin 2)-deficient congenital muscular dystrophy. *FEBS Lett,* 1997. 415(1): p. 33-9.
[43] Angelin, A., et al., Mitochondrial dysfunction in the pathogenesis of Ullrich congenital muscular dystrophy and prospective therapy with cyclosporins. *Proc Natl Acad Sci U S A*, 2007. 104(3): p. 991-6.
[44] Engvall, E. and U.M. Wewer, The new frontier in muscular dystrophy research: booster genes. *FASEB J,* 2003. 17(12): p. 1579-84.
[45] Qiao, C., et al., Amelioration of laminin-alpha2-deficient congenital muscular dystrophy by somatic gene transfer of miniagrin. *Proc Natl Acad Sci U S A,* 2005. 102(34): p. 11999-2004.
[46] Rooney, J.E., P.B. Gurpur, and D.J. Burkin, Laminin-111 protein therapy prevents muscle disease in the mdx mouse model for Duchenne muscular dystrophy. *Proc Natl Acad Sci U S A*, 2009. 106(19): p. 7991-6.
[47] Bentzinger, C.F., et al., Overexpression of mini-agrin in skeletal muscle increases muscle integrity and regenerative capacity in laminin-alpha2-deficient mice. *FASEB J,* 2005. 19(8): p. 934-42.
[48] Gawlik, K.I., et al., Laminin alpha1 chain improves laminin alpha2 chain deficient peripheral neuropathy. *Hum Mol Genet,* 2006. 15(18): p. 2690-700.
[49] Gawlik, K.I. and M. Durbeej, Transgenic overexpression of laminin alpha1 chain in laminin alpha2 chain-deficient mice rescues the disease throughout the lifespan. *Muscle Nerve,* 2010. 42(1): p. 30-7.

[50] Gawlik, K., et al., Laminin alpha1 chain reduces muscular dystrophy in laminin alpha2 chain deficient mice. *Hum Mol Genet,* 2004. 13(16): p. 1775-84.

[51] Moll, J., et al., An agrin minigene rescues dystrophic symptoms in a mouse model for congenital muscular dystrophy. *Nature*, 2001. 413(6853): p. 302-7.

[52] Meinen, S., et al., Linker molecules between laminins and dystroglycan ameliorate laminin-alpha2-deficient muscular dystrophy at all disease stages. *J Cell Biol,* 2007. 176(7): p. 979-93.

[53] Gawlik, K.I., et al., Distinct roles for laminin globular domains in laminin alpha1 chain mediated rescue of murine laminin alpha2 chain deficiency. *PLoS One*, 2010. 5(7): p. e11549.

[54] Rooney, J.E., et al., Laminin-111 protein therapy reduces muscle pathology and improves viability of a mouse model of merosin-deficient congenital muscular dystrophy. *Am J Pathol,* 2012. 180(4): p. 1593-602.

[55] Vachon, P.H., et al., Merosin and laminin in myogenesis; specific requirement for merosin in myotube stability and survival. *J Cell Biol,* 1996. 134(6): p. 1483-97.

[56] Petrof, B.J., et al., Dystrophin protects the sarcolemma from stresses developed during muscle contraction. *Proc Natl Acad Sci U S A,* 1993. 90(8): p. 3710-4.

[57] Hall, T.E., et al., The zebrafish candyfloss mutant implicates extracellular matrix adhesion failure in laminin alpha2-deficient congenital muscular dystrophy. *Proc Natl Acad Sci U S A,* 2007. 104(17): p. 7092-7.

[58] Girgenrath, M., et al., Inhibition of apoptosis improves outcome in a model of congenital muscular dystrophy. *J Clin Invest,* 2004. 114(11): p. 1635-9.

[59] Dominov, J.A., et al., Muscle-specific BCL2 expression ameliorates muscle disease in laminin {alpha}2-deficient, but not in dystrophin-deficient, mice. Hum Mol Genet, 2005. 14(8): p. 1029-40.

[60] Girgenrath, M., et al., Pathology is alleviated by doxycycline in a laminin-alpha2-null model of congenital muscular dystrophy. *Ann Neurol*, 2009. 65(1): p. 47-56.

[61] Erb, M., et al., Omigapil ameliorates the pathology of muscle dystrophy caused by laminin-alpha2 deficiency. *J Pharmacol Exp Ther,* 2009. 331(3): p. 787-95.

[62] Menetrey, J., et al., Growth factors improve muscle healing in vivo. *J Bone Joint Surg Br,* 2000. 82(1): p. 131-7.

[63] Fukushima, K., et al., The use of an antifibrosis agent to improve muscle recovery after laceration. *Am J Sports Med,* 2001. 29(4): p. 394-402.

[64] Lefaucheur, J.P. and A. Sebille, Basic fibroblast growth factor promotes in vivo muscle regeneration in murine muscular dystrophy. *Neurosci Lett, 1*995. 202(1-2): p. 121-4.

[65] Lefaucheur, J.P. and A. Sebille, Muscle regeneration following injury can be modified in vivo by immune neutralization of basic fibroblast growth factor, transforming growth factor beta 1 or insulin-like growth factor I. *J Neuroimmunol*, 1995. 57(1-2): p. 85-91.

[66] Lavasani, M., et al., Nerve growth factor improves the muscle regeneration capacity of muscle stem cells in dystrophic muscle. *Hum Gene Ther,* 2006. 17(2): p. 180-92.

[67] Sandri, M., Signaling in muscle atrophy and hypertrophy. *Physiology* (Bethesda), 2008. 23: p. 160-70.

[68] Barton, E.R., et al., Muscle-specific expression of insulin-like growth factor I counters muscle decline in mdx mice. *J Cell Biol,* 2002. 157(1): p. 137-48.

[69] Rommel, C., et al., Mediation of IGF-1-induced skeletal myotube hypertrophy by PI(3)K/Akt/mTOR and PI(3)K/Akt/GSK3 pathways. *Nat Cell Biol*, 2001. 3(11): p. 1009-13.
[70] Palazzolo, I., et al., Overexpression of IGF-1 in muscle attenuates disease in a mouse model of spinal and bulbar muscular atrophy. *Neuron*, 2009. 63(3): p. 316-28.
[71] Lynch, G.S., et al., IGF-I treatment improves the functional properties of fast- and slow-twitch skeletal muscles from dystrophic mice. *Neuromuscul Disord,* 2001. 11(3): p. 260-8.
[72] Hayes, A. and D.A. Williams, Examining potential drug therapies for muscular dystrophy utilising the dy/dy mouse: I. Clenbuterol. *J Neurol Sci,* 1998. 157(2): p. 122-8.
[73] Sneddon, A.A., et al., Elevated IGF-II mRNA and phosphorylation of 4E-BP1 and p70(S6k) in muscle showing clenbuterol-induced anabolism. *Am J Physiol Endocrinol Metab,* 2001. 281(4): p. E676-82.
[74] Kline, W.O., et al., Rapamycin inhibits the growth and muscle-sparing effects of clenbuterol. *J Appl Physiol,* 2007. 102(2): p. 740-7.
[75] Bodine, S.C., et al., Akt/mTOR pathway is a crucial regulator of skeletal muscle hypertrophy and can prevent muscle atrophy in vivo. *Nat Cell Biol*, 2001. 3(11): p. 1014-9.
[76] Kumar, A., et al., Muscle-specific expression of insulin-like growth factor 1 improves outcome in Lama2Dy-w mice, a model for congenital muscular dystrophy type 1A. *Hum Mol Genet,* 2011. 20(12): p. 2333-43.
[77] Hayashi, Y.K., et al., Massive muscle cell degeneration in the early stage of merosin-deficient congenital muscular dystrophy. *Neuromuscul Disord,* 2001. 11(4): p. 350-9.
[78] Rutkowski, A., et al., Report on the myomatrix conference april 22-24, 2012, university of nevada, reno, nevada, USA. *Neuromuscul Disord,* 2013. 23(2): p. 188-91.
[79] Li, Q. and I.M. Verma, NF-kappaB regulation in the immune system. *Nat Rev Immunol*, 2002. 2(10): p. 725-34.
[80] Bakkar, N. and D.C. Guttridge, NF-kappaB signaling: a tale of two pathways in skeletal myogenesis. *Physiol Rev*, 2010. 90(2): p. 495-511.
[81] Mourkioti, F. and N. Rosenthal, NF-kappaB signaling in skeletal muscle: prospects for intervention in muscle diseases. *J Mol Med,* 2008. 86(7): p. 747-59.
[82] Costelli, P., et al., Muscle protein waste in tumor-bearing rats is effectively antagonized by a beta 2-adrenergic agonist (clenbuterol). Role of the ATP-ubiquitin-dependent proteolytic pathway. *J Clin Invest,* 1995. 95(5): p. 2367-72.
[83] Strassmann, G., et al., Mechanisms of experimental cancer cachexia. Local involvement of IL-1 in colon-26 tumor. *J Immunol*, 1993. 150(6): p. 2341-5.
[84] Reid, M.B. and J.S. Moylan, Beyond Atrophy: Redox Mechanisms of Muscle Dysfunction in Chronic Inflammatory Disease. *J Physiol*, 2011.
[85] Acharyya, S., et al., Interplay of IKK/NF-kappaB signaling in macrophages and myofibers promotes muscle degeneration in Duchenne muscular dystrophy. *J Clin Invest,* 2007. 117(4): p. 889-901.
[86] Taniguchi, M., et al., Expression profiling of muscles from Fukuyama-type congenital muscular dystrophy and laminin-alpha 2 deficient congenital muscular dystrophy; is congenital muscular dystrophy a primary fibrotic disease? *Biochem Biophys Res Commun*, 2006. 342(2): p. 489-502.

[87] Noguchi, S., et al., cDNA microarray analysis of individual Duchenne muscular dystrophy patients. *Hum Mol Genet*, 2003. 12(6): p. 595-600.

[88] MacDonald, E.M. and R.D. Cohn, TGFbeta signaling: its role in fibrosis formation and myopathies. *Curr Opin Rheumatol,* 2012. 24(6): p. 628-34.

[89] Aoki, F. and I. Kojima, Therapeutic potential of follistatin to promote tissue regeneration and prevent tissue fibrosis. *Endocr J,* 2007. 54(6): p. 849-54.

[90] Khalil, N., et al., TGF-beta 1, but not TGF-beta 2 or TGF-beta 3, is differentially present in epithelial cells of advanced pulmonary fibrosis: an immunohistochemical study. *Am J Respir Cell Mol Biol,* 1996. 14(2): p. 131-8.

[91] Isaka, Y., et al., Gene therapy by skeletal muscle expression of decorin prevents fibrotic disease in rat kidney. *Nat Med,* 1996. 2(4): p. 418-23.

[92] Nevo, Y., et al., Fibrosis inhibition and muscle histopathology improvement in laminin-alpha2-deficient mice. *Muscle Nerve*, 2010. 42(2): p. 218-29.

[93] Habashi, J.P., et al., Losartan, an AT1 antagonist, prevents aortic aneurysm in a mouse model of Marfan syndrome. *Science*, 2006. 312(5770): p. 117-21.

[94] Elbaz, M., et al., Losartan, a therapeutic candidate in congenital muscular dystrophy: studies in the dy(2J) /dy(2J) mouse. *Ann Neurol*, 2012. 71(5): p. 699-708.

[95] Meinen, S., S. Lin, and M.A. Ruegg, Angiotensin II type 1 receptor antagonists alleviate muscle pathology in the mouse model for laminin-alpha2-deficient congenital muscular dystrophy (MDC1A). *Skelet Muscle*, 2012. 2(1): p. 18.

[96] Bodine, S.C., et al., Identification of ubiquitin ligases required for skeletal muscle atrophy. *Science*, 2001. 294(5547): p. 1704-8.

[97] Gomes, M.D., et al., Atrogin-1, a muscle-specific F-box protein highly expressed during muscle atrophy. *Proc Natl Acad Sci U S A*, 2001. 98(25): p. 14440-5.

[98] Carmignac, V., R. Quere, and M. Durbeej, Proteasome inhibition improves the muscle of laminin alpha2 chain-deficient mice. *Hum Mol Genet,* 2011. 20(3): p. 541-52.

[99] Lin, N.Y., et al., Autophagy regulates TNFalpha-mediated joint destruction in experimental arthritis. *Ann Rheum Dis*, 2012.

[100] Carmignac, V., et al., Autophagy is increased in laminin alpha2 chain-deficient muscle and its inhibition improves muscle morphology in a mouse model of MDC1A. *Hum Mol Genet*, 2011. 20(24): p. 4891-902.

[101] Meinen, S., et al., Apoptosis inhibitors and mini-agrin have additive benefits in congenital muscular dystrophy mice. *EMBO Mol Med*, 2011. 3(8): p. 465-79.

In: Laminins: Structure, Biological Activity and Role in Disease ISBN: 978-1-62618-568-5
Editors: Demetrius C. Adams and Emil O. Garcia

Chapter 6

Roles of Laminins in Epithelial Morphogenesis in the Liver

Naoki Tanimizu[1,*], Yamato Kikkawa[2] and Toshihiro Mitaka[1]
[1]Department of Tissue Development and Regeneration,
Research Institute for Frontier Medicine, Sapporo
Medical University School of Medicine
[2]Laboratory of Clinical Biochemistry, Tokyo University
of Pharmacy and Life Science, Japan

Abstract

The liver performs a wide range of metabolic functions, produces serum proteins, and protects the body from infection. To perform these physiological functions efficiently, epithelial cells in the liver, namely hepatocytes and cholangiocytes, establish three-dimensional (3D) tissue structures consisting of hepatic cords and bile ducts. It has been demonstrated that extracellular matrix proteins, especially laminins, play important roles in differentiation of liver stem/progenitor cells and in the formation of tissue structures. Liver progenitor cells acquire metabolic functions similar to hepatocytes and form bile canalicular-like structures *in vitro* in the presence of Matrigel, which contains α1-containing laminin. In 3D culture, liver progenitor cells differentiate along the cholangiocyte lineage and form cysts with the central lumen, a process that depends on α1-containing laminin.

Recently, we identified that two laminin isoforms have distinct roles during bile duct development; α1-containing laminin is favorable for inducing cholangiocyte lineage at the initial step of bile duct development, whereas α5-containing laminin is essential for the completion of ductular structure formation. Thus, liver epithelial cells sequentially utilize α1- and α5-containing laminins to drive distinctive processes during bile duct morphogenesis. In this review, we describe the roles of laminins in the morphogenesis of

* All correspondence should be addressed to Naoki Tanimizu, Ph.D., Department of Tissue Development and Regeneration, Research Institute for Frontier Medicine, Sapporo Medical University School of Medicine, S-1, W-17, Chuo-ku, Sapporo 060-8556, JAPAN. E-mail: tanimizu@sapmed.ac.jp.

hepatocytes and cholangiocytes, in the maintenance of liver stem/progenitor cells, and in pathogenesis.

Introduction

In epithelial tissues, the basal domains of polarized cells face towards the basal lamina, which consists of extracellular matrix (ECM) proteins, such as laminins, nidogen/entactin, and type IV collagen. Laminins (heterotrimers of α, β, and γ chains) play pivotal roles in the development and maintenance of epithelial tissues (Durbeej, 2010). There are 19 different laminin isoforms consisting of 5 α, 3 β, and 3 γ chains. Expression of these isoforms is tightly regulated in a spatiotemporal manner during development; therefore, each isoform is assumed to have distinctive functions. However, the specific laminin isoforms that are responsible for transmitting signals to epithelial cells during the stepwise process of epithelial morphogenesis remain largely unknown.

The liver contains two types of epithelial cells, hepatocytes and cholangiocytes. Hepatocytes are the liver parenchymal cells that perform metabolic functions. Cholangiocytes form bile ducts that connect the liver to the intestine to drain the bile produced by hepatocytes, and secrete water and ions to modulate bile composition. In addition, cholangiocytes are associated with the basal lamina throughout bile duct development, suggesting that ECM proteins, including laminins, contribute to the differentiation and morphogenesis of cholangiocytes. Although the laminin subunits that surround bile ducts in an adult liver are well established (Kikkawa et al., 2005), the laminin isoforms contributing in processes of bile duct development are largely unknown.

This review summarizes recent findings regarding the role of laminins in liver development, regeneration, and pathogenesis. In particular, we describe the roles of laminins in bile duct development, regulation of hepatocyte morphogenesis, as well as proliferation and differentiation of liver stem/progenitor cells. Furthermore, we discuss the possibility that laminins affect the adhesive property of sinusoidal endothelial cells (SECs) during liver regeneration and that of hepatocellular carcinoma (HCC) cells.

Laminins in the Early Stage of Liver Development

The definitive foregut endoderm, from which liver bud emerges, is highly associated with basal lamina, which is identified by a layer of laminin. Although the basal lamina may be important for formation of the gut tube, it is degraded when hepatic progenitor cells spread out into the transversum mesenchyme during expansion of the liver primordium. During this process, Prox1 may be important for regulating the ECM protein turnover by the upregulation of matrix metalloproteinase (MMP)-2 (Papoutsi et al., 2007). Actually, it was reported that the liver did not grow into the surrounding mesenchymal tissues in Prox1 knockout (KO) mice, in which the basal lamina was intact around the liver bud (Sosa-Pineda et al., 2000). Beyond this stage, the basal lamina underlies the mesothelium and exists in the portal triad containing the portal vein, hepatic artery, and bile ducts.

Roles of Laminins in the Development of Bile Ducts

In midgestation (around E14–15 in mice), fetal liver stem/progenitor cells called hepatoblasts split into hepatocytes and cholangiocytes. Cholangiocytes form tubular structures called bile ducts via multiple morphogenic processes (Figure 1). The hepatoblasts adjacent to the portal vein are committed to becoming cholangiocytes and form a single cell layer called the ductal plate (Figs. 1B and E). The Notch and TGFβ signals are major signaling pathways inducing cholangiocyte lineage from hepatoblasts.

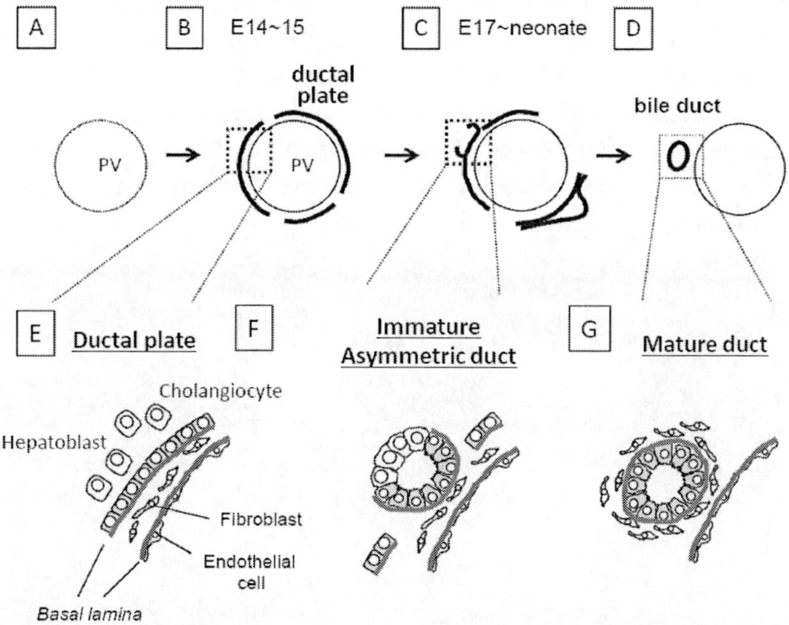

Figure 1. Deposition of laminin during the development of bile ducts. Near the portal vein, hepatoblasts differentiate into cholangiocytes and form the ductal plate. Ductal plates are reorganized into bile duct tubules through immature "asymmetric" duct structures, which consist of cholangiocytes and hepatoblasts. According to the model proposed by Antoniou et al. and Zong et al., hepatoblasts in asymmetric ducts differentiate into cholangiocytes and subsequently form mature duct structures. During tubular morphogenesis, the basal lamina contains α1- and α5-containing laminins which are produced by periportal fibroblasts and cholangiocytes, respectively. The amount of α1-containing laminin gradually decreases and subsequently disappears in the basal lamina that surrounds mature ducts.

The ductal plate is reorganized into tubular structures called bile ducts. During bile duct formation, immature "asymmetric" duct structures are first formed; the lumen is surrounded by both hepatoblasts and cholangiocytes (Figures 1C and F). In the current model proposed by Antoniou et al. (Antoniou et al., 2009), hepatoblasts on the parenchymal side eventually differentiate into cholangiocytes, and subsequently form mature "symmetric" duct structures (Figs. 1D and G). In addition to Notch and TGFβ signals, transcription factors such as Sox9, HNF1β, and Hes1 have been implicated in the process (Antoniou et al., 2009; Coffinier et al., 2002; Sumazaki et al., 2004; Zong et al., 2009). The laminin layer is already visible on the basal side of cholangiocytes at the ductal plate stage. Since laminins exist adjacent to cholangiocytes throughout life, they are assumed to be essential for the development and maintenance of bile ducts.

To study this process, we established a 3D culture system using the liver progenitor cell line named hepatic progenitors proliferating on laminin (HPPL), which was derived from Dlk-1$^+$ hepatoblasts (Tanimizu et al., 2007; Tanimizu et al., 2004a). In the presence of Matrigel, we observed that HPPL polarized and formed cysts with a central lumen. Moreover, we demonstrated that α1-containing laminin (laminin 111), a major component of Matrigel, transmitted signals via α6β1 integrin for the formation of these cysts (Figure 2). However, we also observed that the exogenous α1-containing laminin was dispensable for the maintenance of the cyst structure (Tanimizu et al., 2012). RT-PCR analysis indicated that HPPL expressed laminin α5, β2, and β1, and immunofluorescent analysis demonstrated that the layer of α5-containing laminin was deposited on the basal side of the cyst structure. Thus, while exogenous α1-containing laminin supports cyst formation, endogenous α5-containing laminin contributes to the maintenance of the cyst structure. Based on these *in vitro* studies, we assumed that α1- and α5-containing laminins have distinctive roles in cholangiocyte differentiation and morphogenesis.

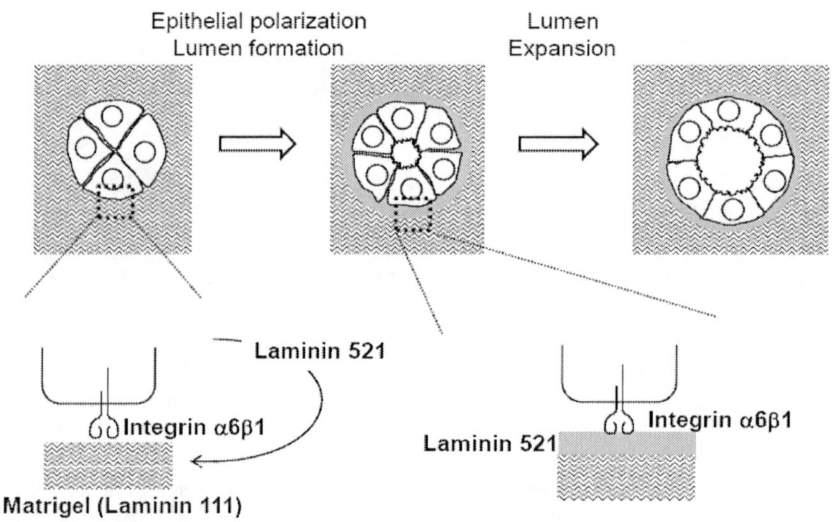

Figure 2. Roles of exogenous α1-containing laminin and endogenous α5-containing laminin in cystogenesis of HPPL. HPPL receive signals from α1-containing laminin (laminin 111) via α6β1 integrin to establish apico–basal polarity and to form cysts. In addition, HPPL produce α5-containing laminin (laminin 521), which is important for the structural maintenance of cysts and the enlargement of the central lumen.

Laminins containing α1 and α5 chains are observed on the basal side of the ductal plate *in vivo*, although the laminin layer next to certain cholangiocytes is devoid of the α5 chain. Subsequently, α1-containing laminin gradually disappears and α5-containing laminin remains adjacent to mature bile ducts.

We observed that p75NTR$^+$ cells, including periportal fibroblasts, produce α1-containing laminin whereas cholangiocytes are responsible for the production of α5-containing laminin

(Tanimizu et al., 2012). Therefore, this laminin isoform switch is achieved through altered laminin production by these cell types.

Time-specific expression of α1-containing laminin suggests that it plays an important role in hepatoblast lineage specification. We cultured Dlk1$^+$ hepatoblasts isolated from E14.5 fetal liver on α1- or α5-containing laminin and stimulated them with TGFβ. We observed that TGFβ induced cholangiocyte markers including Sox9, EpCAM, and cytokeratin 19 (CK19), in hepatoblasts more efficiently on α1-containing laminin than on α5-containing one (Tanimizu et al., 2012) (Figure 3). In addition, we activated the Notch signal in hepatoblasts by overexpressing the Notch intracellular domain (NICD) and observed that similar to TGFβ, the Notch signal induced Sox9 more efficiently on α1-containing laminin. Moreover, on α1-containing laminin, the Sox9 expression was increased further through simultaneous activation of both pathways (TGFβ and Notch) in comparison with individual activation of each signal. These results suggest that α1-containing laminin favors the activation of signals that induce the differentiation of cholangiocytes from hepatoblasts.

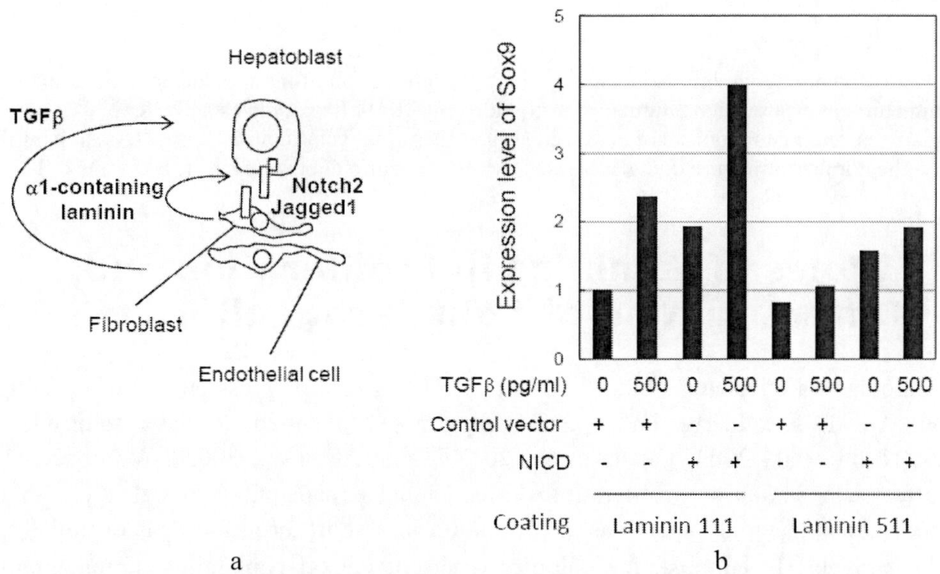

Figure 3. Induction of cholangiocyte markers by TGFβ and Notch signals on α1- and α5-containing laminin. Notch and TGFβ signaling pathways are activated in hepatoblasts near the portal vein and induce the cholangiocyte lineage. TGFβ is supplied from endothelial cells and fibroblasts, whereas the Notch signal is activated by the interaction between Jagged1$^+$ periportal fibroblasts and Notch2$^+$ hepatoblasts. α1-containing laminin is supplied from periportal fibroblasts (A). Dlk-1$^+$ hepatoblasts were cultured on dishes coated with laminin 111 or laminin 511, and subsequently stimulated by the addition of TGFβ and/or by introducing a retrovirus vector coding the Notch intracellular domain (NICD). Sox9 expression was further increased by activation of both TGFβ and Notch signals on laminin 111.

To examine the roles of laminins containing α1 and α5 chains in bile duct morphogenesis, we analyzed duct structures in E17 livers of laminin α5 KO mice (Tanimizu et al., 2012). A similar number of CK19$^+$ cholangiocytes emerged around the PV in control and mutant livers, which is consistent with the notion that α1-containing laminin is sufficient for the induction of cholangiocyte markers by TGFβ and Notch signals *in vitro*. However, the number of mature duct structures decreased in the mutant livers, while the number of immature "asymmetric ducts" increased (Figure 4). These results indicated that the absence of α5-containing laminin delayed or attenuated bile duct morphogenesis. Taken together, α1-containg laminin was sufficient for inducing the cholangiocyte lineage from hepatoblasts in response to induction signals, whereas α5-containing laminin was necessary to complete the formation of mature duct structure.

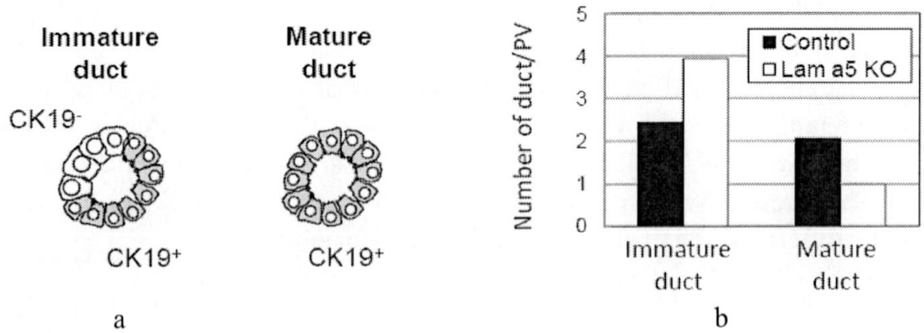

Figure 4. α5-Containing laminin is indispensable for the formation of mature duct structures in the liver. Immature ducts, in which a lumen is surrounded by $CK19^-$ hepatoblasts and $CK19^+$ cholangiocytes, and mature ducts are observed around the portal vein in an E17 liver (A). In laminin α5 KO liver, the number of mature ducts decreased, while immature ducts increased (B).

Roles of Laminins in Proliferation and Differentiation of Stem/Progenitor Cells

Hepatoblasts are isolated from E14.5 fetal liver based on expression of surface antigens, including Dlk-1, E-cadherin, and CD13, and have been shown to have colony-forming abilities (Nitou et al., 2002; Okada et al., 2012; Tanimizu et al., 2003). We observed that ECM affected hepatoblasts proliferation during clonal expansion. Although type I and IV collagens supported rapid hepatoblast proliferation in a short term, the cells eventually died within 1–2 weeks. In contrast, the colonies emerging on α1-containing laminin gradually expanded over a longer period, and the cells cultured on α1-containing laminin were capable of proliferation even after passages. HPPL were established from the colonies expanded on α1-containing laminin (Tanimizu et al., 2004a). HPPL were capable of bidirectional differentiation as hepatoblasts, differentiating into hepatocytes in the presence of OSM and Matrigel and cholangiocytes in 3D culture. HPPL expressed integrin α3, α6, β1, and β4 chains. Therefore, we assumed that α3β1, α6β1, and α6β4 integrins would be responsible for transmitting signals from α1-containing laminin. Recently, we observed that neonatal liver stem/progenitor cells could expand on α1-containing laminin while keeping their original marker expression and the ability to differentiate into mature hepatocytes. Thus, α1-containing laminin contributes to the self-renewal of liver stem/progenitor cells *in vitro*.

Hepatoblasts isolated from E14.5 mouse livers can differentiate into hepatocytes in the presence of dexamethasone and OSM (Kamiya et al., 1999). This hepatocytic differentiation is further augmented in the presence of Matrigel (Kamiya et al., 2002). It was proposed that this might be regulated by α1-containing laminin in the Matrigel through β1 integrin (Kamiya et al., 2002). In addition to gene expression related to differentiated hepatic functions, morphological changes, such as formation of bile canalicular-like structures, have been observed in the presence of Matrigel. Similar effects of Matrigel on hepatocytic differentiation have been demonstrated in adult liver progenitor cells (Sugimoto et al., 2002) and liver progenitor cell lines (unpublished data). On the other hand, basal lamina has not been observed adjacent to hepatocytes in normal liver tissues. Therefore, it remains unknown whether Matrigel ectopically promotes hepatocytic differentiation *in vitro* or whether laminins exist at undetectable levels near hepatocytes, which might be sufficient to support hepatocyte differentiation *in vivo*.

Roles of Laminins in Pathophysiological Conditions of Adult Livers

ECM proteins, such as type I, III, and IV collagens and fibronectin have been observed in the parenchymal region of the normal adult liver (Baptista et al., 2011), along with chondroitin and heparin sulfate proteoglycans (Wang et al., 2011). These ECM proteins might collectively function to induce and/or maintain the differentiation status of mature hepatocytes (MHs) (Wang et al., 2011). However, laminins are hardly detected in the normal liver and the basal lamina cannot be found in the space of Disse, which is the interstitial space between MHs and SECs. On the other hand, the basal lamina is apparent in humans and rodents prior to liver fibrosis (Martinez-Hernandez and Martinez, 1991; Xu et al., 2003). This process called "capillarization" is defined by the loss of SEC fenestration and the development of basal lamina in the space of Disse (Schaffner and Poper, 1963). Although the basal lamina is not developed in the parenchymal region, laminins are transiently induced during liver regeneration (Kikkawa et al., 2005).

In our studies, we demonstrated that α1-containing laminin could be induced in regenerating liver after partial hepatectomy (PHx). Although we could not detect α1-containing laminin in normal liver, it was induced in regenerating livers 3 days after PHx, and the gene expression level increased by 6 days. Hepatic stellate cells (HSCs) are regarded as a major cellular source for ECMs in normal and fibrotic livers, and α1-containing laminin was produced by HSCs in regenerating livers. Hepatocytes isolated from normal and regenerating livers showed similar binding capacity to α1-containing laminin. On the other hand, SECs isolated from regenerating livers could attach and spread on α1-containing laminin, whereas SECs isolated from normal livers could not. Neither HSCs nor Kupffer cells isolated from normal and regenerating livers attached to α1-containing laminin. These results suggest that α1-containning laminin might specifically affect the cellular behavior of SECs in regenerating livers. During liver regeneration after PHx, replication of hepatocytes is prominent during the first 2–3 days. Subsequently, liver tissue architecture, including the sinusoids, is reconstructed 7–10 days after PHx. Considering that α1-containing laminin is expressed between days 3 and 10 after PHx, it is possible that α1-containing laminin regulates the migration of SECs during reconstitution of the sinusoids. In addition to α1-containing laminin, α4-containing laminin, which is observed adjacent to the portal vein and the hepatic artery in the normal liver, increases in sinusoids of regenerating livers. However, hepatocytes, SECs, and HSCs isolated from regenerating livers did not attach to dishes coated with α4-containing laminin (Kikkawa et al., 2005). Therefore, the role of α4-containing laminin in liver regeneration remains unknown. A close association of liver stem/progenitors with laminins was reported in normal and injured mouse models, in which intrinsic stem/progenitor cells were activated (Reid et al., 1992). Although MHs undergo self-duplication to compensate for the MH loss after acute liver injury, stem/progenitor cells are also activated to facilitate tissue repair in chronically injured livers or when hepatocyte replication is blocked. It is known that in regenerating rat livers, which are treated with 2-acetylaminofluorene to block hepatocyte replication prior to PHx, adult liver stem/progenitors (known as oval cells) are activated and expanded (Oh et al., 2002). In this model, oval cells expressing Dlk-1 (a marker for hepatoblasts), which might be committed to hepatocytes, partially lost their association with the laminin layer (Tanimizu et al., 2004b). Recently, it was demonstrated that osteopontin$^+$ (OPN) liver stem/progenitor cells, which were induced during chronic hepatic injury caused by a choline-deficient ethionine-supplemented (CDE) diet, could differentiate into MHs (Espanol-Suner et al., 2012). Espanol-Suner et al. labeled these stem/progenitor cells using the expression of a yellow fluorescent protein (YFP) in OPN-CreERT2/ROSA26$^{YFP/YFP}$ mice after injecting tamoxifen. Feeding these mice with the CDE diet for 3 weeks followed by normal diet for 2 weeks, YFP expression was observed in 2.5% of hepatocytes, indicating that they were derived from OPN$^+$ stem/progenitor cells. Comparing YFP$^+$ stem/progenitors and YFP$^+$ hepatocytes, they observed that laminins did not surround YFP$^+$ hepatocytes, suggesting that ECM, including laminins, may regulate hepatocytic differentiation of stem/progenitor cells. When expression of ECM proteins was inhibited using Iloprost (a synthetic analogue of prostaglandin I_2 known to block TGFβ-mediated fibrogenesis), the differentiation of OPN$^+$ stem/progenitor cells into hepatocytes was remarkably promoted in mice being fed with the CDE diet. Thus, the association of stem/progenitor cells with laminins may also contribute to the maintenance of stem/progenitor cells *in vivo*. HCC is a devastating disease with poor prognosis. We examined the expression of laminins in human HCCs using specific antibodies to laminin chains (Kikkawa et al., 2008). In the livers from patients with HCC, α5-containing laminin was only observed in the region of HCC, regardless of the tumor grade. In addition, we observed that α5-containing laminin was detected in the culture media of all tested HCC cell lines, such as HepG2, HLE, HLF, and HuH-7. These results indicate that HCC cells may ectopically produce laminins. In addition, we observed that HCC cells could readily attach to α5-containing laminin (laminin 511) but not to α1-containing laminin (laminin 111) *in vitro*. An inhibition assay using a recombinant protein and blocking antibodies indicated that HCC cells bound to α5-containing laminin via α1β1/α3β1/α6β1 integrins and Lutheran/Basal cell adhesion molecule (Lu/B-CAM). Taken together, our results suggest that HCC cells establish a tumor microenvironment suitable for their expansion by producing α5-containing laminin.

Perspective

The switch from α1- to α5-containing laminin during development is widely observed in many organs, and recent studies have identified the specific roles of laminin isoforms. In the submandibular gland, $\alpha 5$-containing laminin signals strengthen FGF signals by increasing receptor expression during morphogenesis (Rebustini et al., 2007). In addition, Kikkawa and Miner generated laminin $\alpha 5$ KO mice expressing a mutant laminin $\alpha 5$ chain, in which the entire $\alpha 5$ LG domain was replaced by $\alpha 1$ LG. Using this model, they observed that the mutant laminin $\alpha 5$ chain failed to rescue the developmental defects observed in $\alpha 5$ KO mice, including digit septation, placental labyrinth morphology, lung lobe septation, and vascularization of kidney glomeruli (Kikkawa and Miner, 2006). Furthermore, Gao et al. described that $\alpha 1$-containing laminin failed to rescue defects in hair follicle morphogenesis in laminin $\alpha 5$ KO mice (Gao et al., 2008). We demonstrated that liver epithelial cells interacted with $\alpha 1$-containing laminin during the establishment of apico–basal polarity and the apical lumen and that they utilized $\alpha 5$-containing laminin to further advance bile duct morphogenesis and maintain bile duct structure. Moreover, we have observed the possibility that FGFs promote lumen formation in 3D culture of HPPL (unpublished data). Further analysis is required to understand the precise mechanisms by which laminin isoforms affect morphogenesis of hepatic bile ducts and epithelial morphogenesis in other organs/tissues.

In the adult liver, in addition to $\alpha 5$-containing laminin, $\alpha 3$- and $\alpha 4$-containing laminins exist around bile ducts (Kikkawa et al., 2005). Laminin $\alpha 3$ is known to be a component of hemidesmosomes. Laminin $\alpha 3$ KO mice demonstrated severe skin blistering (Ryan et al., 1999) and abnormal glomerulogenesis (Abrass et al., 2006), and died within 3 days after birth. $\alpha 6\beta 4$ integrin, a major receptor for $\alpha 3$-containing laminin, is expressed on cholangiocytes. Although we have examined effects of anti-$\beta 4$ integrin antibody on HPPL cyst formation and maintenance, cyst structure was not disrupted in the presence of the antibody. Given that defects in bile duct development have not been reported in laminin $\alpha 3$ KO mice, and that $\beta 4$ integrin expression arises later than $\beta 1$ integrin (Tanimizu et al., 2012), the laminin $\alpha 3$–$\beta 4$ integrin signaling axis might be important for the maturation/maintenance of bile ducts rather than their development.

Although a laminin-rich environment induces HPPL to form luminal structures in 3D culture, HPPL do not form tubular structures similar to bile ducts *in vivo*. This is true even with exposure to high HGF concentrations, which is often used to induce tubular morphogenesis in cell lines derived from the kidney such as MDCK and IMCD cells (Zegers et al., 2003). In contrast, when HPPL cysts were transferred into collagen gel, they formed some extended structures (unpublished data), suggesting the possibility that a crosstalk between signals from laminins and collagens is involved in tubular morphogenesis. At present, we are seeking culture conditions for inducing tubular morphogenesis in HPPL to study the detailed molecular mechanisms governing bile duct morphogenesis *in vitro*.

Generating KO mice is an extremely powerful technique to show the importance of target genes in developmental processes. However, it is challenging to continuously follow the progression of these morphogenic processes *in vivo* to reveal underlying molecular mechanisms. 3D culture systems have provided a way to recapitulate many aspects of *in vivo* tissue morphogenesis. Furthermore, using live cell imaging, it is possible to observe epithelial morphogenesis under a microscope. Therefore, to observe the relationships between morphogenesis and ECM proteins, we are currently establishing HPPL that express nidogen tagged with a fluorescence protein and YFP-actin. In conclusion, future studies involving these *in vivo* and *in vitro* models should help in shedding light on the roles of laminins in epithelial differentiation and morphogenesis.

References

Abrass, C.K., Berfield, A.K., Ryan, M.C., Carter, W.G., and Hansen, K.M. (2006). Abnormal development of glomerular endothelial and mesangial cells in mice with targeted disruption of the lama3 gene. *Kidney Int.* 70, 1062-1071.

Antoniou, A., Raynaud, P., Cordi, S., Zong, Y., Tronche, F., Stanger, B.Z., Jacquemin, P., Pierreux, C.E., Clotman, F., and Lemaigre, F.P. (2009). Intrahepatic bile ducts develop according to a new mode of tubulogenesis regulated by the transcription factor SOX9. *Gastroenterology* 136, 2325-2333.

Baptista, P.M., Siddiqui, M.M., Lozier, G., Rodriguez, S.R., Atala, A., and Soker, S. (2011). The use of whole organ decellularization for the generation of a vascularized liver organoid. *Hepatology* 53, 604-617.

Coffinier, C., Gresh, L., Fiette, L., Tronche, F., Schutz, G., Babinet, C., Pontoglio, M., Yaniv, M., and Barra, J. (2002). Bile system morphogenesis defects and liver dysfunction upon targeted deletion of HNF1beta. *Development* 129, 1829-1838.

Durbeej, M. (2010). Laminins. *Cell Tissue Res*. 339, 259-268.

Espanol-Suner, R., Carpentier, R., Van Hul, N., Legry, V., Achouri, Y., Cordi, S., Jacquemin, P., Lemaigre, F., and Leclercq, I.A. (2012). Liver progenitor cells yield functional hepatocytes in response to chronic liver injury in mice. *Gastroenterology* 143, 1564-1575 e1567.

Gao, J., DeRouen, M.C., Chen, C.H., Nguyen, M., Nguyen, N.T., Ido, H., Harada, K., Sekiguchi, K., Morgan, B.A., Miner, J.H., et al. (2008). Laminin-511 is an epithelial message promoting dermal papilla development and function during early hair morphogenesis. *Genes Dev*. 22, 2111-2124.

Kamiya, A., Kinoshita, T., Ito, Y., Matsui, T., Morikawa, Y., Senba, E., Nakashima, K., Taga, T., Yoshida, K., Kishimoto, T., et al. (1999). Fetal liver development requires a paracrine action of oncostatin M through the gp130 signal transducer. *EMBO J.* 18, 2127-2136.

Kamiya, A., Kojima, N., Kinoshita, T., Sakai, Y., and Miyaijma, A. (2002). Maturation of fetal hepatocytes in vitro by extracellular matrices and oncostatin M: induction of tryptophan oxygenase. *Hepatology* 35, 1351-1359.

Kikkawa, Y., and Miner, J.H. (2006). Molecular dissection of laminin alpha 5 in vivo reveals separable domain-specific roles in embryonic development and kidney function. *Dev. Biol.* 296, 265-277.

Kikkawa, Y., Mochizuki, Y., Miner, J.H., and Mitaka, T. (2005). Transient expression of laminin alpha1 chain in regenerating murine liver: restricted localization of laminin chains and nidogen-1. *Exp. Cell Res*. 305, 99-109.

Kikkawa, Y., Sudo, R., Kon, J., Mizuguchi, T., Nomizu, M., Hirata, K., and Mitaka, T. (2008). Laminin alpha 5 mediates ectopic adhesion of hepatocellular carcinoma through integrins and/or Lutheran/basal cell adhesion molecule. *Exp. Cell Res.* 314, 2579-2590.

Martinez-Hernandez, A., and Martinez, J. (1991). The role of capillarization in hepatic failure: studies in carbon tetrachloride-induced cirrhosis. *Hepatology* 14, 864-874.

Nitou, M., Sugiyama, Y., Ishikawa, K., and Shiojiri, N. (2002). Purification of fetal mouse hepatoblasts by magnetic beads coated with monoclonal anti-e-cadherin antibodies and their in vitro culture. *Exp. Cell Res.* 279, 330-343.

Oh, S.H., Hatch, H.M., and Petersen, B.E. (2002). Hepatic oval 'stem' cell in liver regeneration. *Semin. Cell Dev. Biol.* 13, 405-409.

Okada, K., Kamiya, A., Ito, K., Yanagida, A., Ito, H., Kondou, H., Nishina, H., and Nakauchi, H. (2012). Prospective isolation and characterization of bipotent progenitor cells in early mouse liver development. *Stem Cells Dev.* 21, 1124-1133.

Papoutsi, M., Dudas, J., Becker, J., Tripodi, M., Opitz, L., Ramadori, G., and Wilting, J. (2007). Gene regulation by homeobox transcription factor Prox1 in murine hepatoblasts. *Cell Tissue Res.* 330, 209-220.

Rebustini, I.T., Patel, V.N., Stewart, J.S., Layvey, A., Georges-Labouesse, E., Miner, J.H., and Hoffman, M.P. (2007). Laminin alpha5 is necessary for submandibular gland epithelial morphogenesis and influences FGFR expression through beta1 integrin signaling. *Dev. Biol.* 308, 15-29.

Reid, L.M., Fiorino, A.S., Sigal, S.H., Brill, S., and Holst, P.A. (1992). Extracellular matrix gradients in the space of Disse: relevance to liver biology. *Hepatology* 15, 1198-1203.

Ryan, M.C., Lee, K., Miyashita, Y., and Carter, W.G. (1999). Targeted disruption of the LAMA3 gene in mice reveals abnormalities in survival and late stage differentiation of epithelial cells. *J. Cell Biol.* 145, 1309-1323.

Schaffner, F., and Poper, H. (1963). Capillarization of hepatic sinusoids in man. *Gastroenterology* 44, 239-242.

Sosa-Pineda, B., Wigle, J.T., and Oliver, G. (2000). Hepatocyte migration during liver development requires Prox1. *Nat. Genet.* 25, 254-255.

Sugimoto, S., Mitaka, T., Ikeda, S., Harada, K., Ikai, I., Yamaoka, Y., and Mochizuki, Y. (2002). Morphological changes induced by extracellular matrix are correlated with maturation of rat small hepatocytes. *J. Cell Biochem.* 87, 16-28.

Sumazaki, R., Shiojiri, N., Isoyama, S., Masu, M., Keino-Masu, K., Osawa, M., Nakauchi, H., Kageyama, R., and Matsui, A. (2004). Conversion of biliary system to pancreatic tissue in Hes1-deficient mice. *Nat. Genet.* 36, 83-87.

Tanimizu, N., Kikkawa, Y., Mitaka, T., and Miyajima, A. (2012). α1- and α5-containing laminins regulate the development of bile ducts via β1 integrin signals. *J. Biol. Chem.* 287, 28586-28597.

Tanimizu, N., Miyajima, A., and Mostov, K.E. (2007). Liver progenitor cells develop cholangiocyte-type epithelial polarity in three-dimensional culture. *Mol. Biol. Cell* 18, 1472-1479.

Tanimizu, N., Nishikawa, M., Saito, H., Tsujimura, T., and Miyajima, A. (2003). Isolation of hepatoblasts based on the expression of Dlk/Pref-1. *J. Cell Sci.* 116, 1775-1786.

Tanimizu, N., Saito, H., Mostov, K., and Miyajima, A. (2004a). Long-term culture of hepatic progenitors derived from mouse Dlk+ hepatoblasts. *J. Cell Sci.* 117, 6425-6434.

Tanimizu, N., Tsujimura, T., Takahide, K., Kodama, T., Nakamura, K., and Miyajima, A. (2004b). Expression of Dlk/Pref-1 defines a subpopulation in the oval cell compartment of rat liver. *Gene Expr. Patterns* 5, 209-218.

Wang, Y., Cui, C.B., Yamauchi, M., Miguez, P., Roach, M., Malavarca, R., Costello, M.J., Cardinale, V., Wauthier, E., Barbier, C., et al. (2011). Lineage restriction of human hepatic stem cells to mature fates is made efficient by tissue-specific biomatrix scaffolds. *Hepatology* 53, 293-305.

Xu, B., Broome, U., Uzunel, M., Nava, S., Ge, X., Kumagai-Braesch, M., Hultenby, K., Christensson, B., Ericzon, B.G., Holgersson, J., et al. (2003). Capillarization of hepatic sinusoid by liver endothelial cell-reactive autoantibodies in patients with cirrhosis and chronic hepatitis. *Am. J. Pathol.* 163, 1275-1289.

Zegers, M.M., O'Brien, L.E., Yu, W., Datta, A., and Mostov, K.E. (2003). Epithelial polarity and tubulogenesis in vitro. *Trends Cell Biol.* 13, 169-176.

Zong, Y., Panikkar, A., Xu, J., Antoniou, A., Raynaud, P., Lemaigre, F., and Stanger, B.Z. (2009). Notch signaling controls liver development by regulating biliary differentiation. *Development* 136, 1727-1739.

In: Laminins: Structure, Biological Activity and Role in Disease ISBN: 978-1-62618-568-5
Editors: Demetrius C. Adams and Emil O. Garcia

Chapter 7

Role of Laminin on Uterine Remodeling and Embryo Development during Pregnancy

Bruno Zavan, Andréa Mollica do Amarante-Paffaro [*]
and Valdemar Antonio Paffaro Junior
Integrative Animal Biology Laboratory – LABAInt, Biomedical Science Institute,
UNIFAL-MG, Alfenas, MG, Brazil

Abstract

The Laminin – which is made up of three polypeptide chains, α, β and y, joined by bisulfide bonds – interacts with entactin and heparin sulfate to assist the cells in binding to Extracellular Matrix, and therefore participate in cell proliferation and differentiation process. Pregnancy is a phenomenon that, to be viable, depends on major events including the implantation process, trophoblast cells invasion, as well as placental and embryo development. For that reason, the uterus undergoes major morphological changes during pregnancy and depends on the laminin support to ensure the gestation progress. In the non-pregnant mouse uterus, the laminin can be found in the endometrium region exclusively as a component of surface epithelium, glands and capillaries basal membranes, whereas the endometrial stroma is negative. After successful mating the blastocyst apposition and adhesion to the uterine luminal epithelium are assisted by many molecules including laminin. Subsequently, the decidual cells start to synthesize laminin, as well as others molecules, and it is deposited in the endometrial Extracellular Matrix at the time of trophoblast invasion. The integrin repertoire expressed by trophoblasts cells indicate that the laminin and fibronectin support trophoblast invasion until it reaches the maternal spiral arteries. The proper invasion of trophoblast cells is critical to ensure the correct development of the placenta and its protection from maternal immune system rejection, and consequently, it's important to guarantee nutrients, gases, and maternal protective antibodies perfusion into the fetal circulation, producing hormones that control pregnancy, and contributing to the maintenance of a healthy gestation. The role laminin

[*] Email: paffaroj@unifal-mg.edu.br.

in pregnancy is not restricted to the implantation process, trophoblast invasion and placental development. The major role of this molecule is to support the cellular remodeling that occurs in embryo development process. The absence of any laminin polypeptide chain (α, β and y) is sufficient to cause multiple embryonic defects including syndactyly, failure of neural tube closure, renal agenesis/failure of glomerulogenesis, placental vessel malformations, defects of the neuromuscular junction/glomerulus, skeletal muscle/peripheral nerve, and microvasculature/peripheral nerve. These sets of data demonstrate that laminin is imperative for both uterine remodeling and embryo development during gestation.

Introduction

The integrity of epithelium, blood vessel and various gland membranes is supported by two fundamental molecules: Laminin and collagen IV (DerRest, 1991). The Laminin is made up of three polypeptide chains, α, β and y, joined by disulfide bonds (Colognato and Yurchenco, 2000). This molecule interacts with entactin and heparin sulfate that assist the cells to bind to Extracellular Matrix which participate in cell proliferation and differentiation process (Farrrar and Carson, 1992). Accordingly, laminin is important in various organs and tissues involved in many physiological events as one of the constituent molecules of the basal lamina and thus as part of a layer that serves as extracellular binding substrate for anchorage or cell migration.

Pregnancy is a phenomenon that, to be viable, depends on major events including the implantation process, trophoblast cells invasion, as well as placental and embryo development. Therefore, the uterus undergoes major morphological changes during pregnancy, and this chapter intends to point out the role of laminin during this period, as well as its importance in supporting pregnancy progress.

Laminin and the Uterine Remodeling during Pregnancy

To study the reproduction biology, rats and mice are commonly used as animal model according to the great similarity between these hemochorial placentation organisms. At the mouse not-pregnant uterus, the laminin can be found at endometrium region exclusively as a component of basement membranes of surface epithelium, glands and capillaries, whereas the endometrial stroma is negative (Karkavelas et al. 1998). The laminin can also be found at myometrium in vascular and smooth muscle basement membranes (Karkavelas et al. 1998).

After successful mating, the oocyte fertilized develops to morula stage while moving in the uterine tube lumen until reaches the uterine cavity where the implantation may occur. In mice, the implantation occurs between gestations day 4-6 (Abrahamsohn and Zorn 1993), and this event can be described as a process starting with the apposition and adhesion of a competent blastocyst to the apical surface of the endometrial epithelium, and this process is also called as "attachment. The attachment is only possible if the uterine luminal epithelium is receptive, which in turn occurs just in a short period of time termed "implantation window" (Psychoyos, 1976, 1986; Finn, 1977).

During blastocyst implantation, the trophoblast cells rapid penetrate at endometrium, and this process can be considered similar to the invasion of basement membranes and stroma by tumor cells (Yagel et al., 1988; Graham and Lala, 1992). The cancer cells migrate into tissues and vessels probably involving a series of cell-matrix interactions. Attachment primarily to laminin or fibronectin is considered to be required for traction and stabilization of the migrating cell (Liotta et al., 1983). As the Basal Membrane of endometrial epithelium is rich in laminin and fibronectin (Karkavelas et al. 1998), it is plausible to suppose that trophoblast cells bind to laminin and fibronectin before invade the endometrial stroma.

According to Clark and colleagues (1993), laminin appears in the primary decidual zone on the gestational day 6 in rats, and can be seen in the epitelial basal membrane, vessels and gland epitheliaum on the gestational day 7 (Clark et al., 1993).

Laminin which is found exclusively in the basement membrane is synthesized by decidual cells, as well as fibronectin and collagen type IV, and it is deposited in the endometrial extracellular matrix at the time of trophoblast invasion (Wewer et al. 1985; Kisalus et al. 1987) working as a multifunctional protein influencing complex biological processes like cell adhesion, growth, differentiation, and migration (Kleinmann et al. 1985). Therefore the laminin expression appears to accompany cellular differentiation with its increased expression during implantation (Korgun et al. 2007).

After the implantation process, the receptive endometrium undergoes to an extension invasion by the trophoblast cells throughout the early pregnancy period, until reach the access of maternal spiral arteries, and replace the endothelium of the arterial wall forming endovascular-trophoblast (Loke and King, 1995; Kaufmann and Scheffen 1992). At this time, the invasive trophoblast cells no longer express the integrin $\alpha6\beta4$, on the other hand it gain the expression of another kind of integrin, the $\alpha1\beta1$ integrin for binding with the collagen and laminins, besides the expression of $\alpha5\beta1$ that binds to fibronectin present in the decidual extracellular matrix (Damsky et al. 1992; Earl et al. 1990). The expression of $\alpha1\beta1$ integrin seems to be spatio-temporal regulated, since, the capacity of trophoblast invasion is different in first trimester of pregnancy compared to term placenta in humans (Damsky et al. 1994), as it can be confirmed *in vitro* from studies about the invasive ability of trophoblast cells isolated from these two different periods of pregnancy (Librach et al. 1991; Zhou et al. 1993).

Laminin is the essential compound of the deciadualized uterine stroma, and can increase trophoblast invasion capacity as it's expression is upregulated during this process (Burrows et al., 1996, Zhang et al. 2000), and this molecule seems to be present predominantly at the antimesometrial side of embryo implantation site (Zavan et al. 2010).

As observed in goats (Guillomot 1999), laminin and collagen IV have been reported to disappear from the uterine laminal epithelium after blastocyst implantation and trophoblast invasion to the stroma, or else to be present in limited quantities in mice. It has been suggested that this down regulation is associated with decidual cell differentiation (Blankenship and Given, 1995). Some studies (Gürgen et al. 2012) attribute the laminin decreasing to basal membrane destruction in order to facilitate implantation. Threfore the laminin are able to support the successful maintenance of strong secretion, implantation and invasion in epithelium (Gürgen et al. 2012).

Thus, the extracellular compounds, such as laminin, are important to certify precise trophoblast invasion to ensure the correct development of the placenta and its protection from maternal immune system rejection, and also to ensure the transporting nutrients, gases, and maternal protective antibodies into the fetal circulation, producing hormones that control pregnancy, and contributing to the maintenance of a healthy pregnancy.

Posteriorly of these orchestrated process of blastocyst implantation and extravillous trophoblast invasion, a construction of hemochorial placenta starts, and this transient organ is able to feed and protect the growing fetus. The placentation is result from a correct extravillous trophoblast invasion with in turn depends on laminin. Studies have shown that the mutation of *Lama5 (alpha 5 laminin)* causes defects on placentation process (Miner et al. 2004).

The role of laminin in pregnancy is not restricted to the implantation process, trophoblast invasion and placental development. The major role of this molecule is to support the cellular remodeling that occurs in embryo development process.

Laminin and the Embryo Development

Extracellular Matrix proteins can be found in the very early stage of mammalian development. Laminin was firstly detected in the 8-cell stage of the mouse embryo and then shown in Reichert's Membrane that is a yolk sac membrane of the developing embryo (Korgun et al. 2007).

The first differentiated cell type to arise from the Inner Cell Mass during the mouse peri-implantation development is the primitive endoderm (Li et al., 2003; Rossant, 2004). Endoderm cells, on the other hand, secrete laminin, type IV collagen and other basement membrane components. The visceral and parietal endoderm originate from primitive endoderm, and after implantation, the visceral-endoderm-covered epiblast cells adjacent to the basement membrane polarize to form a pseudostratified columnar epithelium, which is precursor of three definitive germ layers. The absence of cell-contact with the basement membrane results in cell death by apoptosis, leading to proamniotic cavity formation. Laminin is required for epiblast polarization, embryoid bodies cavitation, and protection of adhered cells from apoptosis (Murray and Edgar 2000).

These first laminins to be expressed in mammalian embryos are Laminins-111 and 511 (a5b1g1) which are essential for basement membrane assembly and early embryonic morphogenesis (Smyth et al. 1999; Huang et al. 2003; Miner et al. 2004;Urbano et al. 2009). Therefore, the laminin seems important to germ cells migration in order to promote germ layers formation, and interestingly the absence of laminin is also important to ensure the proamniotic cavity formation, during peri-implantation period.

The significance of laminin in this process was confirmed by studies (Smyth et al., 1999) showing that ablation of the laminin $\gamma 1$ chain in mice abolishes Basal Membrane formation and also arrests the development at the peri-implantation stage with a phenotype similar to that of integrin-$\beta 1$-null mice, suggesting that laminin and $\alpha\beta 1$ integrins interaction is essential for early embryogenesis (Smyth et al. 1999; Liu et al. 2009).

For assembling the extraembryonic Reichert's Membrane the laminin a1 seems to be required whereas basement membrane formation of the embryonic plate requires either the Lma1 or Lma5 subunit (Miner et al. 2004). The laminin subunit widely expressed is the laminin-a5. The inactivation of Laminin-a5 gene results in multiple embryonic defects including syndactyly, failure of neural tube closure, renal agenesis/failure of glomerulogenesis, and placental vessel malformations (Miner et al. 1998). In turn, inactivation of the genes coding for the laminin b2, a2 and a4 subunits, which are expressed later in a more tissue-restricted manner caused defects of the neuromuscular junction/glomerulus, skeletal muscle/peripheral nerve, and microvasculature/peripheral nerve, respectively, with survival into the post-natal period (Noakes et al. 1995a; Noakes et al. 1995b; Miyagoe et al. 1997; Thyboll et al. 2002; Wallquist et al. 2005).

During the mammalian brain neurogenesis, neuroectodermally derived glia become elongated bipolar cells that extend from the ventricular (apical) to the pial (basal) surface, the latter contacting the pial basement membrane through glial endfeet. These cells can give rise to neurons and affect their fate. Of note, the basal cell processes of these radial glial cells appear to act as a guide for the migration of newly formed neurons in the ventricular zone to the pia, leading to the formation of cell layers (Kosodo and Huttner 2009). Mutations affecting expression of laminins, the nidogen-binding site in the laminin g1 subunit (Lmg1DLEb3), perlecan, a6 and b1-integrins, stroglycan, focal adhesion kinase, and integrin-linked kinase have been found to result in separation of glial cells from the basement membrane because of apparent loss of ECM integrity, and defects of neuronal cell migration and cell layer formation leading to cobblestone-type lissencephaly (Georges-Labouesse et al. 1998; Costell et al. 1999; Graus-Porta et al. 2001; Halfter et al. 2002; Moore et al. 2002; Beggs et al. 2003; Niewmierzycka et al. 2005).

This set of data decrypted on the literature demonstrate unequivocally that these extracellular molecule called laminin is imperative for both uterine remodeling and embryo development during gestation and further studies of this molecule could clarify new important aspects of the normal and pathological pregnancy.

References

Abrahamsohn PA, Zorn TM. Implantation and decidualization in rodents. *J Exp Zool.* 1993 Sep 1;266(6):603-28.

Beggs HE, Schahin-Reed D, Zang K, Goebbels S, Nave KA, Gorski J, Jones KR, Sretavan D, Reichardt LF. 2003. FAK deficiency in cells contributing to the basal lamina results in cortical abnormalities resembling congenital muscular dystrophies. Neuron 40: 501–514.

Blankenship TN, Given RL. Loss of laminin and type IV collagen in uterine luminal epithelial basement membranes during blastocyst implantation in the mouse. *Anat Rec.* 1995 Sep;243(1):27-36.

Burrows, T.D., King, A., Loke, Y.W., 1996. Trophoblast migration during human placental implantaton. *Hum. Reprod.* 2, 307–321.

Clark DE, Hurst PR, McLennan IS et al. (1993) Immunolocalization of collagen type I and laminin in the uterus on days 5 to 8 of embryo implantation in the rat. *Anat Rec* 237:8–20

Colognato, H. and Yurchenco, P. D. (2000). Form and function: the laminin family of heterotrimers. *Dev. Dyn.* 218, 213-234.

Costell M, Gustafsson E, Aszódi A, Mörgelin M, Bloch W, Hunziker E, Addicks K, Timpl R, Fässler R. Perlecan maintains the integrity of cartilage and some basement membranes. *J Cell Biol.* 1999 Nov 29;147(5):1109-22.

Damsky CH, Fitzgerald ML, Fisher SJ. Distribution patterns of extracellular matrix components and adhesion receptors are intricately modulated during first trimester cytotrophoblast differentiation along the invasive pathway, in vivo. *J Clin Invest* 1992; 89: 210-22.

Damsky CH, Librach C, Lim KH, Fitzgerald ML, McMaster MT, Janatpour M, Zhou Y, Logan SK, Fisher SJ. Integrin switching regulates normal trophoblast invasion. *Development* 1994; 120: 3657-66.

DerRest, M.V., Garrone, R., 1991. Collagen family of proteins. *FASEB J.* 5, 2814–2823.

Earl U, Estlin C, Bulmer JN. Fibronectin and laminin in the early human placenta. *Placenta* 1990; 11: 223-31.

Farrrar, J.D., Carson, D.D., 1992. Differential temporal and spatial expression of mRNA encoding extracellular matrix component in decidua during the periimplantation period. *Biol. Reprod.* 46, 1095–1108.

Finn, C.A. (1977) The implantation reaction. In Wynn, R.M. (eds), *Biology of the Uterus*. Plenum Press, New York, pp. 245-303.

Georges-Labouesse E, Mark M, Messaddeq N, Gansmüller A. Essential role of alpha 6 integrins in cortical and retinal lamination. *Curr Biol*. 1998 Aug 27;8(17):983-6.

Graham CH, Lala PK. Mechanisms of placental invasion of the uterus and their control. *Biochem Cell Biol*. 1992 Oct-Nov;70(10-11):867-74.

Graus-Porta D, Blaess S, Senften M, Littlewood-Evans A, Damsky C, Huang Z, Orban P, Klein R, Schittny JC, Müller U. Beta1-class integrins regulate the development of laminae and folia in the cerebral and cerebellar cortex. *Neuron*. 2001 Aug 16;31(3):367-79.

Guillomot M. Changes in extracellular matrix components and cytokeratins in the endometrium during goat implantation. *Placenta*. 1999 May;20(4):339-45.

Gürgen SG, Erdoğan D, Coşkun ZK, Cansu A. The effect of valproic acid and oxcarbazepine on the distribution of adhesion molecules in embryo implantation. *Toxicology*. 2012 Feb 26;292(2-3):71-7.

Halfter W, Dong S, Yip YP, Willem M, Mayer U. A critical function of the pial basement membrane in cortical histogenesis. *J Neurosci*. 2002 Jul 15;22(14):6029-40.

Huang CC, Hall DH, Hedgecock EM, Kao G, Karantza V, Vogel BE,Hutter H, Chisholm AD, Yurchenco PD,Wadsworth WG. 2003. Laminin a subunits and their role in C. elegans development. *Development* 130: 3343–3358.

Karkavelas, G., Kefalides, N. A., Amenta, P. S., Martinez-Hernandez, A. Comparative Ultrastructural Localization of Collagen Types III, IV, VI and Laminin in Rat Uterus and Kidney. *Journal of ultrastucture and Molecular Structure Research.* 100, 137-155. 1988.

Kaufmann P, Scheffen I. Placental development. In: Polin RA, Fox WW, editors. Fetal and neonatal physiology. Philadelphia: W.B. Saunders; 1992. p. 47-56.

Kisalus LL, Herr JC, Little CD (1987) Immunolocalization of extracellular matrix proteins and collagen synthesis in firsttrimester human decidua. *Anat Rec* 218:402–415

Kleinmann HK, Cannon FB, Laurie GW et al. (1985) Biological activities of laminin. *J Cell Biochem* 27:317–325

Korgun ET, Cayli S, Asar M, Demir R. Distribution of laminin, vimentin and desmin in the rat uterus during initial stages of implantation. *J Mol Histol*. 2007 Aug;38(4):253-60.

Korgun ET, Cayli S, Asar M, Demir R. Distribution of laminin, vimentin and desmin in the rat uterus during initial stages of implantation. *J Mol Histol*. 2007 Aug;38(4):253-60.

Kosodo Y, Huttner WB. Basal process and cell divisions of neural progenitors in the developing brain. *Dev Growth Differ*. 2009 Apr;51(3):251-61.

Li, S., Edgar, D., Fassler, R., Wadsworth, W. and Yurchenco, P. D. (2003). The role of laminin in embryonic cell polarization and tissue organization. *Dev*. Cell 4, 613-624.

Librach CL, Werb Z, Fitzgerald ML, Chiu K, Corwin NM, Esteves RA, Grobelny D, Galardy R, Damsky CH, Fisher SJ. 92-kD type IV collagenase mediates invasion of human cytotrophoblasts. *J Cell Biol* 1991; 113: 437-49.

Liotta LA, Rao CN, Barsky SH. Tumor invasion and the extracellular matrix. *Lab Invest*. 1983 Dec;49(6):636-49.

Loke YW, King A. Human trophoblast development. In: Human implantation: cell biology and immunology. Cambridge: Cambridge University Press; 1995.

Miner JH, Cunningham J, Sanes JR. 1998. Roles for laminin in embryogenesis: Exencephaly, syndactyly, and placentopathy in mice lacking the laminin a5 chain. *J Cell Biol,* 143: 1713–1723.

Miner JH, Li C,Mudd JL, Go G, Sutherland AE. 2004. Compositional and structural requirements for laminin and basement membranes during mouse embryo implantation and gastrulation. *Development* 131: 2247–2256.

Miyagoe Y, Hanaoka K, Nonaka I, Hayasaka M, Nabeshima Y, Arahata K, Nabeshima Y, Takeda S. 1997. Laminin a2 chain-null mutant mice by targeted disruption of the Lama2 gene: a new model of merosin (laminin 2)-deficient congenital muscular dystrophy. *FEBS Lett.* 415: 33–39.

Moore SA, Saito F, Chen J, Michele DE, HenryMD,Messing A, Cohn RD, Ross-Barta SE, Westra S, Williamson RA, et al. 2002. Deletion of brain dystroglycan recapitulates aspects of congenital muscular dystrophy. *Nature* 418: 422–425.

Murray P, Edgar D. Regulation of programmed cell death by basement membranes in embryonic development. *J Cell Biol.* 2000 Sep 4;150(5):1215-21.

Niewmierzycka A, Mills J, St-Arnaud R, Dedhar S, Reichardt LF. 2005. Integrin-linked kinase deletion from mouse cortex results in cortical lamination defects resembling cobblestone lissencephaly. *J Neurosci* 25: 7022–7031.

Noakes PG, Gautam M,Mudd J, Sanes JR,Merlie JP. 1995a. Aberrant differentiation of neuromuscular junctions in mice lacking s- laminin/laminin b2. *Nature* 374: 258–262.

Noakes PG, Miner JH, Gautam M, Cunningham JM, Sanes JR, Merlie JP. 1995b. The renal glomerulus of mice lacking s-laminin/laminin b2: Nephrosis despite molecular compensation by laminin b1. *Nat Genet* 10: 400–406.

Psychoyos, A. (1976) Hormonal control of uterine receptivity for nidation. *J. Reprod. Fertil.,* 25 (Suppl.), 17-28.

Psychoyos, A. (1986) Uterine receptivity for nidation. *Ann. N.Y. Acad. Sci.,* 476, 36-42.

Rossant, J. (2004). Lineage development and polar asymmetries in the peri-implantation mouse blastocyst. *Semin. Cell Dev. Biol.* 15, 573-581.

Smyth N, Vatansever HS,Murray P,MeyerM, Frie C, Paulsson M, Edgar D. 1999. Absence of basement membranes after targeting the LAMC1 gene results in embryonic lethality due to failure of endoderm differentiation. *J Cell Biol* 144: 151–160.

Thyboll J, Kortesmaa J, Cao R, Soininen R,Wang L, Iivanainen A, Sorokin L, Risling M, Cao Y, Tryggvason K. 2002. Deletion of the laminin a4 chain leads to impaired microvessel maturation. *Mol Cell Biol* 22: 1194–1202.

Urbano JM, Torgler CN, Molnar C, Tepass U, Lopez-Varea A, Brown NH, de Celis JF, Martin-Bermudo MD. 2009. Drosophila laminins act as key regulators of basement membrane assembly and morphogenesis. *Development* 136: 4165–4176.

Wallquist W, Plantman S, Thams S, Thyboll J, Kortesmaa J, Lannergren J, Domogatskaya A, Ogren SO, Risling M, Hammarberg H, et al. 2005. Impeded interaction among Schwann cells and axons in the absence of laminin alpha4. *J Neurosci* 25: 3692–3700.

Wewer UM, Faber M, Liotta LA et al. (1985) Immunochemical and ultrastructural assessment of the nature of the pericellular basement membrane of human decidual cells. *Lab Invest* 53:624–633

Yagel S, Parhar RS, Jeffrey JJ, Lala PK. Normal nonmetastatic human trophoblast cells share in vitro invasive properties of malignant cells. *J Cell Physiol.* 1988 Sep;136(3):455-62.

Zavan B, Paffaro AM, Joazeiro PP, Yamada AT, Paffaro VA Jr. Immunocytochemical studies of adhesion molecules on mouse UNK cells and their extracellular matrix ligands during mouse pregnancy. *Anat Rec* (Hoboken). 2010 Jun;293(6):1081-8.

Zhang C, Duan E, Cao Y et al. (2000) Effect of 32/67 kDa lamininbinding protein antibody on mouse embryo implantation. *J Reprod Fertil* 119:137–142

Zhou Y, Damsky CH, Chiu K, Roberts JM, Fisher SJ. Preeclampsia is associated with abnormal expression of adhesion molecules by invasive cytotrophoblasts. *J Clin Invest* 1993; 91: 950-60.

In: Laminins: Structure, Biological Activity and Role in Disease ISBN: 978-1-62618-568-5
Editors: Demetrius C. Adams and Emil O. Garcia

Chapter 8

Human Gingival Laminin-332 Degradation by Oral Bacteria

Pirjo Pärnänen[1,2,*] and Timo Sorsa[1,2]
[1]Institute of Dentistry, [1]University of Helsinki and Department of Oral and Maxillofacial Diseases, [2]Helsinki University Central Hospital, Finland

Abstract

In this study laminin-332 (LM-332), labelled with ^{35}S- methionine, was isolated from an immortal human oral gingival epithelial cell line. The LM-332 degradation by common oral periodontal bacteria and *Lactobacilli* was detected by SDS-PAGE/ fluorography.

The degradation assays were performed at two different pHs by using bacterial whole cells and 10x concentrated growth media fractions. *Porphyromonas gingivalis* cells at pH 6 and 10x concentrated growth media at pH 6 and 7.3 caused complete and efficient degradation of LM-332 chains. Degradation was also observed as appearance of 30-40 kDa bands by *Prevotella intermedia* cell fraction and *Prevotella nigrescens* 10 x growth media fractions at pH 6.

No marked degradation of LM-332 by either fraction or pH occurred by *Lactobacilli*, *Actinobacillus actinomycetemcomitans*, *Fusobacterium nucleatum, Peptostreptococcus micros* or *Treponema denticola* cells. Degradation of the basement membrane causes the loss of integrity of the epithelial layer, and this may form an invasion route for oral pathogens, especially through the oral gingival junctional epithelium.

Keywords: Gingival laminin; degradation; Oral bacteria

* Correspondence: Pirjo Pärnänen, Institute of Dentistry, University of Helsinki, Finland. Email: pirjo.parnanen@helsinki.fi.

Introduction

Laminins are glycoproteins found in basement membranes underlying the epithelium. They are secreted by epithelial cells and are composed of α-, β- and γ-chains. Each laminin contains a specific combination of these subunits and form heterotrimeric molecules with structural and signaling functions (Aumailley and Smyth, 1998; Jones et al. 2000). Laminin-332 (LM-332), formerly known as laminin-5 (new nomenclature see Aumailley et al. 2005), is a glycoprotein found in the basement membrane of squamous epithelium. Human gingival LM-332 is 460 kDa in size and contains α3,- β3- and γ2-chains. Unprocessed 190 kDa α3 chain is secreted by the epithelial cells and processed to α3' of 160 kDa in size, unprocessed β3 and γ2 are 135 kDa and 155 kDa in size (Takkunen et al. 2006).

Oral epithelium is the first mechanical barrier against pathogen invasion. The structure and integrity of this stratified squamous epithelial layer may be compromised by proteolytic degradation of the proteinaceous components of the epithelial layer. It is known that both host factors or certain enzymes such as matrix metalloproteinases and a broad spectrum of microbial proteases can degrade laminin, fibronectin, collagens and other extracellular and basement membrane matrix components (Singh et al. 2012). Previous laminin degradation studies with oral bacteria are limited to one degradation study, which was performed with undefined laminin (Pike et al. 1996). Human epithelial LM-332 has not been commercially available. The oral microbes form a biofilm on the epithelium. In the mouth the most pathogenic species are periodontal pathogens, such as *Porphyromonas*, *Prevotella* and *Peptostreptococcus* species, and *Aggregatibacter actinomycetemcomitans*. They pose proteases which may cause marked periodontal tissue destruction (Feng and Weinberg 2000). These bacteria have also been associated with cardiovascular diseases (Buhlin et al. 2011) or may additionally cause septicemia (Parahitiyawa et al. 2009). The oral biofilm contains commensal microbes, lactobacilli, which are considered harmless. However, there are reports of *Lactobacillus* bacteremia (Borriello et al. 2003; Salminen et al. 2002, 2006). With this background the present study wished to test the ability of a set of oral bacterial cell and 10x concentrated growth media fractions to degrade human gingival derived LM-332 which would enable them to penetrate deeper into host tissue.

Materials and Methods

Human LM-332 was isolated by immunoprecipitation from a human spontaneous immortalized mucosal keratinocyte cell line (Pärnänen et al. 2008). Briefly, LM-332 (labelled with ^{35}S-methionine, 25 μCi/ml) was immunoprecipited from the keratinocyte growth medium with polyclonal rabbit antiserum against LM-332 and bound to sepharose-beads.

L. rhamnosus GG, *L. rhamnosus* LC 705, *L. rhamnosus* 5.5 a, *L. casei* shirota, *L. casei* 921, *L. johnsoni* La1, *L. paracasei* 8.16b, *L. paracasei* 8.12a, *L. rhamnosus* 11.4a, *L. reuterii* SD2112 were grown in MRS- broth (de Man Rogosa- Sharpe) for 8 hrs at 37 °C and centrifuged for 10 min at 4 °C. The growth media were 10x concentrated by lyophilisation. The cells were washed twice with 50 mM Tris-HCl; 0.2M NaCl; 1 mM $CaCl_2$; pH 7,6 (TNC-buffer), centrifugated at 14000 rpm, 4 °C for 10 min and then disrupted by sonication on ice (Branson 250 sonicator, Branson, Danbury, CT, USA). The periodontopathogens were grown as given by Jie Bao et al. (2008). *Porphyromonas gingivalis* ATCC 33277 (*P.g.*), *Prevotella nigrescens* ATCC 33563 (*P.n*), *Peptostreptococcus micros* ATCC 33270 (*P.m*), *Prevotella intermedia* ATCC 25611 (*P.i*), *Actinobacillus actinomycetemcomitans* (*A.a.*) ATCC 29523, 43718, 33384, 787, 37399 (serotypes a-e) and *Fusobacterium nucleatum* ATCC 25586 (*F.n.*) were grown for 48h. *Actinobacillus actinomyctemcomitans* (*A.a.*) were grown for 6 h. Only the cells of *F.n.* were disrupted, and all the growth medias were prepared by the methods described above. *Treponema denticola* ATCC 35405 (*T.d.*) lyophilisate was diluted to a concentration of 1 mg/ml into PBS pH 7.3. 15 μl of these bacterial cell or growth media fractions were incubated with LM-332 sepharose-beads at 37 °C for 23 h. The degradation was performed at two different pH values: pH 6 and 7.3. Large size reducing 6.5% SDS-PAGE and fluorography was performed according to Pärnänen et al. (2008). 20 μl of ^{14}C labelled protein mixture was used as a molecular weight standard.

Results

Porhyromonas gingivalis cells at pH 6 and 10x concentrated growth media at pH 6 and 7.3 caused total degradation of LM-332 chains. The degradation was observed as appearance of additional bands of 30-40 kDa in size were noted by *Prevotella intermedia* cell fraction and *Prevotella nigrescens* 10x growth media fractions at pH 6. No marked degradation of LM-332 by either fraction or pH occurred by lactobacilli, *Actinobacillus actinomycetemcomitans*, *Fusobacterium nucleatum, Peptostreptococcus micros* or *Treponema denticola* cells.

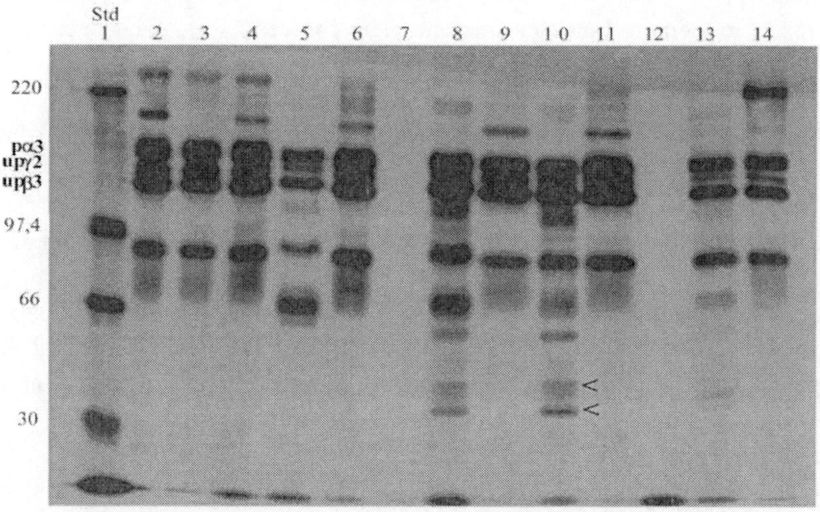

Figure 1. Fluorography. LM-332 degradation at pH 6 by oral bacteria. Lane 1. ^{14}C hot standard (kDa); lane 2. LM-332 control without bacteria; lane 3. LM-332 with PBS, pH 6; lanes 4.-8. bacterial cell fractions: lane 4. *F.n.*; lane 5. *P.n.*; lane 6. *P.m.*; lane 7. *P.g.*; lane 8. *P.i.*; lanes 9.-14. 10 x concentrated growth media: lane. 9. *F.n.*; lane 10. *P.n.*; 11. *P.m.*; lane 12. *P.g.*; lane 13. *P.i.*; lane 14. *T.d.* lyophilisate (1mg/ml). LM-332 chains seen in fluorography: processed α3-chain (pα3, 160 kDa), unprocessed γ2-chain (upγ2, 155 kDa) and unprocessed β3- chain (upβ3, 135 kDa). Degradation products of 30-40 kDa are indicated by arrowheads. For other abbreviations, see the materials and methods section.

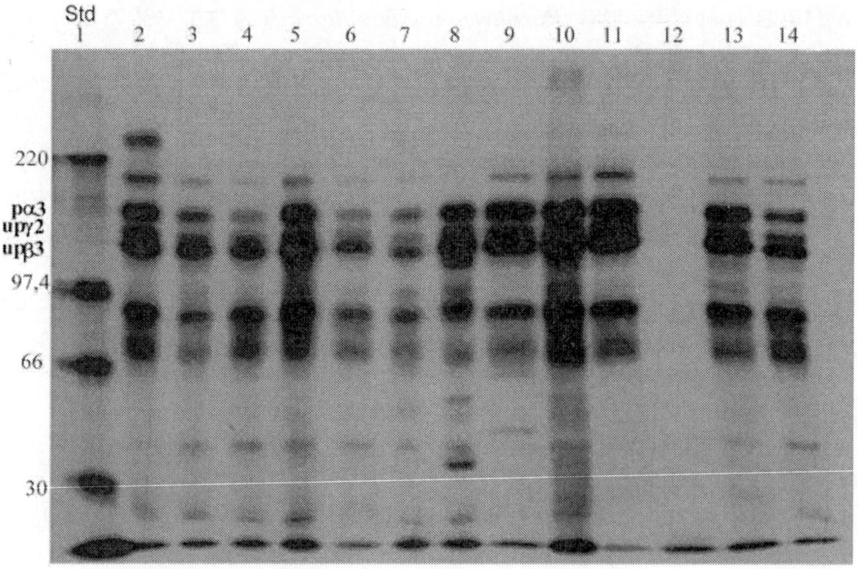

Figure 2. Fluorography. LM-332 degradation at pH 7.3 by oral bacteria. Sample order in lanes as in figure 1.

Conclusion

The periodontopathogen *P.g.* exerted the strongest ability to degrade human epithelial LM-332 in this *in vitro* study. LM-332 degradation was also noted by the *Prevotella* species *P. i.* and *P. n.* to certain but clear extent. This process was pH dependent: LM-332 degradation occurred only at lower pH, with the exception of *P.g.* 10x concentrated growth media which showed degradation at both pH.

Epithelial cell adhesion and mobility is governed by proteolytic processing of the laminin subchains and specific interaction with certain type integrins (Ghosh and Stack, 2000; Tunggal et al. 2000). Processing of γ2- chain is known to be a phased event and epithelial cell stability or enhanced motility depends on the specific cleavage product. β3- chain is thought to be rather stable (Kariya et al. 2012); there are studies indicating that transmembrane serine proteases from prostate cancer cells are able to cleave this chain revealing degradation products of 100 kDa and several of smaller size (Tripathi et al. 2008; 2011). Periodontopathogens exert also serine-like proteases among others (Potempa et al. 2000). The origin of the smaller sized degradation bands of our study will be yet to be verified with immunochemical and mass spectrometric studies. There is a difference in LM-332 structure in the tissue of human and mouse. Mouse LM-332 has two truncated short arms and human LM-332 has only one truncated short arm. This is the result of differential proteolytic processing of the subunits (Jones et al. 2000). This must be considered in the interpretation of LM-332 proteolytic degradation fragment sizes from *in vitro* microbial studies depending on the used substrate source i.e. human or mouse derived.

Periodontopathogenic proteases have been shown to participate in the periodontal tissue destruction process (Sorsa et al. 1992). Uniquely the oral gingival internal basal lamina differs from other parts of the gingiva. The internal basal lamina lies between gingival junctional epithelium and the tooth and contains only LM-332 and it forms a simple extracellular matrix underneath the epithelial cells (Oksanen et al. 2001). It contains no E-cadherin, the major molecule in adherences junctions. These features make the internal basal lamina more vulnerable to microbial proteolytic degradation. The partial or total degradation of the LM-332 chains may cause structural changes in the basement membrane and this may affect epithelial cell motility, cause alterations in signaling processes in the basement membrane zone and obviously enhance periodontitis progression by creating an invasion pathway for oral microbes.

Acknowledgments

In memoriam to Professor of Anatomy Ismo Virtanen (1949-2010). The study was supported by grants from the Helsinki University Central Hospital Research Funds and the Finnish Women Dentists' Association.

Reviewed by professor Jukka H. Meurman; Institute of Dentistry and Helsinki University Central Hospital, Helsinki, Finland.

References

Aumailley, M and Smyth, N. The role of laminins in basement membrane function. *Journal of Anatomy,* 1998, 193, 1-21.

Aumailley, M; Bruckner-Tuderman, L; Carter, WG; Deutzmann, R; Edgar, D; et al. A simplified laminin nomenclature. *Matrix Biology,* 2005, 24, 326-332.

Borriello, SP; Hammes, WP; Holzapfel, W; Marteau, P; Schrezenmeir, M; Vaara, M; Valtonen, V. Safety of probiotic that contain lactobacilli of bifidobacteria. *Clinical Infectious Diseases,* 2003, 36, 775-780.

Buhlin, K; Mäntylä, P; Paju, S; Peltola, JS; Nieminen, MS; Sinisalo, J; Pussinen, PJ. Periodontitis is associated with angiographically verified coronary artery disease. *Journal of Clinical Periodontology,* 2011, 38, 1007-1014.

Feng, Z and Weinberg, A. Role of bacteria in health and disease of periodontal tissues. *Periodontology,* 2000, 40, 50-76.

Ghosh, S. and Stack, MS. Proteolytic modification of laminins: functional consequences. *Microscopy Research and Technique,* 2000, 51, 238-246.

Jie Bao, G; Kari, K; Tervahartiala, T; Sorsa, T; Meurman, JH. Proteolytic activities of oral bacteria on ProMMP-9 and the effect of synthetic proteinase inhibitors. *Open Dental Journal,* 2008, 2, 96-102.

Jones, JCR; Dehart, GW; Gonzales, M; Goldfinger, LE. Laminins: an overview. *Microscopy Research and Technique,* 51, 211-213.

Kariya, Y; Sato, H; Katou, N; Kariya, Y; Miyazaki, K. Polymerized laminin-332 matrix supports rapid and tight adhesion of keratinocytes, suppressing cell migration. *PLoS ONE,* 2012, 7, 1-15.

Oksanen, J; Sorokin, LM; Virtanen, I; Hormia, M. The junctional epithelium around murine teeth differs from gingival epithelium in its basement membrane composition. *Journal of Dental Research,* 2001, 80, 2093-2097.

Parahitiyawa, NB; Jin, LJ; Leung, WK; Yam, WC; Samaranayake, LP. Microbiology of odontogenic bacteremia: beyond endocarditis. *Clinical Microbiology Reviews,* 2009, 22, 46-64.

Pike, RN; Potempa, J; McGraw, W; Coetzer, TH; Travis, J. Characterisation of the binding activities of proteinase-adhesin complexes from Porhyromonas gingivalis. *Journal of Bacteriology,* 1996, 178, 2876-2882.

Potempa, J; Banbula, A; Travis, J. Role of bacterial proteinases in matrix destruction and modulation of host responses. *Periodontology,* 2000, 24, 153-192.

Pärnänen, P; Kari, K; Virtanen, I; Sorsa, T; Meurman, JH. Human laminin-332 degradation by Candida proteinases. *Journal of Oral Pathology and Medicine*, 2008, 6, 329-335.

Salminen, MK; Rautelin, H; Tynkkynen, S; Poussa, T; Saxelin, M; Valtonen, V; Järvinen, A. Lactobacillus bacteremia, species identification, and antimicrobial susceptibility of 85 blood isolates. *Clinical Infectious Disesases,* 2006, 42, e35-e44.

Salminen, MK; Tynkkynen, S; Rautelin, H; Saxelin, M; Vaara, M; Ruutu, P; Sarna, S; Valtonen, V; Järvinen, A. Lactobacillus bacteremia during a rapid increase in probiotic use of Lactobacillus rhamnosus GG in Finland. *Clinical Infectious Diseases,* 2002, 35, 1155-1160.

Singh, B; Fleury, C; Jalalvand, F; Riesbeck, K. Human pathogens utilize host extracellular matrix proteins laminin and collagen for adhesion and invasion of the host. *FEMS Microbiology Reviews,* 2012, 36, 1122-1180.

Sorsa, T; Ingman, T; Suomalainen, K; Haapasalo, M; Konttinen, YT; Lindy, O; Saari, H; Uitto, VJ. Identification of proteases from periodontopathogenic bacteria as activators of latent human neutrophil and fibroblast-type interstitial collagenases. *Infection and Immunity,* 1992, 60, 4491-4495.

Takkunen, M; Grenman, R; Hukkanen, M; Korhonen, M; de Herreros, AG; Virtanen, I. Snail-dependent and –independent epithelial- mesenchymal transition in oral squamous carcinoma cells. *Journal of Histochemistry and Cytochemistry,* 2006, 54, 1263-1275.

Tripathi, M; Nandana, S; Yamashita, H; Ganesan, R; Kirchhofer, D; Quaranta, V. Laminin-332 is a substrate for hepsin, a protease associated with prostate cancer progression. *Journal of Biological Chemistry,* 2008, 283, 30576-30584.

Tripathi, M; Potdar, AA; Yamashita, H; Weidow, B; Cummings, PT; Kirchhofer, D; Quaranta, V. Laminin-332 cleavage by maltriptase alters motility parameters of prostate cancer cells. *The Prostate,* 2011, 71, 184-196.

Tunggal, P; Smyth, N; Paulsson, M; Ott, M-C. Laminins: structure and genetic regulation. *Microscopy Research and Technique,* 2000, 51, 214-227.

In: Laminins: Structure, Biological Activity and Role in Disease ISBN: 978-1-62618-568-5
Editors: Demetrius C. Adams and Emil O. Garcia

Chapter 9

Commentary: Laminins as a Modulator of the Axonal Response to Guidance Cues

Tomoyuki Masuda[1,*], Takashi Shiga[2], Hiroyuki Yaginuma[3] and Shuichi Ueda[1]

[1]Department of Histology and Neurobiology, Dokkyo Medical University School of Medicine, Tochigi, Japan

[2]Department of Neurobiology, University of Tsukuba School of Medicine, Tsukuba, Ibaraki, Japan

[3]Department of Anatomy, Fukushima Medical University School of Medicine, Fukushima, Japan

Abstract

Laminins are extracellular matrix molecules known to be abundant in and around the trajectories of developing axons. In addition to their axonal growth-promoting activity, we showed that the YIGSR peptide of the lamininβ1 (LAMB1) chain modulates the sensory axonal response to chemoattractive cues derived from the embryonic dorsal spinal cord. We further revealed that such modulationis mediated via the 67-kDa receptor for LAMB1 on sensory axons. By employing culture stripe or growth cone turning assays, the two other groups also provided evidence showing that the responses of developing axons to the repellent Slit2 or attractant Netrin-1 are modulated by the laminin peptide YIGSR. Furthermore, those groups also showed that the modulation of cGMP or cAMP levels in the growth cones of developing axons mimics the modulation of the axonal response to the guidance cues. These findings strongly support the idea that extracellular matrix protein laminins have the potential to modulate the response of neuronal growth cones to the chemotrophic guidance molecules by regulatingthe levels of intracellular cyclic nucleotides.

[*] Corresponding author.Tel: +81-282-87-2124, Fax: +81-282-86-1463, E-mail address: tmasu@ dokkyomed.ac.jp.

Keywords: Dorsal root ganglion, Axonal guidance, YIGSR peptide, 67-kDa laminin receptor

Growing axons navigate along specific appropriate pathways to reach their targets. Growth cones of growing axons are guided by the coordinate action of attractive and repulsive guidance cues (reviewed in Tessier-Lavigne and Goodman, 1996; Masuda and Shiga, 2005). Dorsal root ganglion (DRG) sensory neurons extend their axons toward a restricted region of the dorsal spinal cord called the "dorsal root entry zone (DREZ)" at embryonic stage 18-20 (Hamburger and Hamilton, 1992) in the chick embryo (Shiga et al., 2000). Previously, we reported that the dorsal spinal cordsecretes chemoattractive cues for DRG axons to draw these axons toward the DREZ (Masuda et al., 2007). Thus, chemoattractive guidance cues from the embryonic dorsal spinal cord play an important role in DRG axonal guidance. However, none have been identified as acandidate molecule that attracts DRG axons toward the DREZ.

Laminins are heterotrimeric glycoproteins found in the extracellular matrix, each composed of α, β, and γ (LAMA, LAMB, and LAMC) chains (Burgeson et al., 1994). Laminins are known to play a key role in migration, differentiation, and myelination of Schwann cells (Anton et al., 1994; McGarvey et al., 1984; Obremski and Bunge, 1995), and in neuronal growth of a variety of neuronal subtypes *in vitro* (e.g., Adler et al., 1985; Liesiet al., 1989; Manthorpe et al., 1983; Rogers et al., 1983).

The 67-kDa laminin receptor (67LR) is known to be a non-integrin cell-surface receptor for some laminins (reviewed in Ménard et al., 1997). The amino acid sequence of 67LR is extremely well conserved among species, suggesting an essential biological role for this molecule (Gauczynski et al., 2001). 67LR is also known to be a receptor for the YIGSR peptide of the lamininβ1 (LAMB1) chain (e.g., Graf et al., 1987; Massia et al., 1993). In an earlier paper (Masuda et al., 2009), we showed that mesenchymal tissues around DRG axonsin the chick embryoexpress laminin-1 (LM-111) and that DRG neurons display the laminin receptor 67LR. This evidence raised the possibility that LM-111 may be involved in the development of DRG neurons as an axonal growth or axonal guidance molecule. By performing co-culture assays, we demonstrated that soluble LM-111 or the YIGSR peptide of the LAMB1 chain enhances the DRG axonal response to the dorsal spinal cord-derived chemoattraction. Furthermore, we showed thatperturbation of laminin receptor 67LR blocks the promoting effect of the YIGSR peptide, indicating that 67LRs on DRG axons mediate the LM-111-induced modulation of chemoattraction. Additionally, the results of an *in vivo* experiment using the 67LR antibody indicate that 67LR is quite essential for the proper elongation of DRG axons toward the DREZ. Together, both *in vitro* and *in vivo* data strongly suggest that the YIGSR peptide-67LR signals modulatethe DRG axonal response to dorsal spinal cord-derived chemoattractants (Figure 1A).

However, the level of cyclic nucleotides (cAMP, cGMP) in the cytosol of DRG neurons has not yet been examined in the presence of LM-111 or the YIGSR peptide.

According to the correlation between LM-111 and axonal growth or guidance, two other groups have reported that LM-111modifies the axonal response to the chemoattractive guidance molecule Netrin-1 (Höpker et al., 1999) and to the repulsive guidance molecule Slit2 (Nguyen-Ba-Charvet et al., 2001). As described below, this role of LM-111 may be mediated by the YIGSR peptide, as the YIGSR peptide can mimic LM-111-induced modifications of axonal responses (Figure 1B and C; Höpker et al., 1999; Nguyen-Ba-Charvet et al., 2001).

Holt and her colleagues clearly showed that LM-111 converts Netrin-mediated attraction for *Xenopus* neuronal growth cones into repulsion via cell-surface receptor β1 integrins (Höpker et al., 1999). They further showed that the YIGSR peptide, a soluble peptide fragment of LM-111, mimics this LM-111-induced conversion. In addition, they demonstrated that the down-regulation of cyclic AMP in the growth cones lead to the conversion of Netrin-mediated attraction into repulsion.

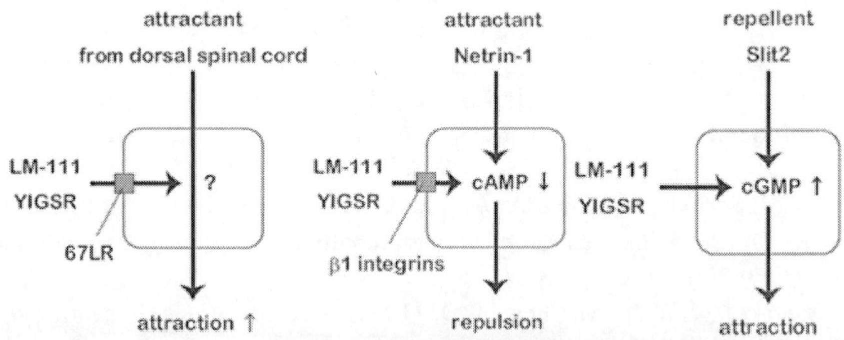

Figure 1. (A) – chick DRG; (B) – Xenopus retinal neuron; (C) – rat DRG. A schematic diagram summarizing three types of LM-111- or YIGSR-mediated modulation of the neuronal response. (A) The cue from the dorsal spinal cord attracts DRG axons. LM-111 or the YIGSR peptide, a component of the LAMB1 chain, around DRG neurons enhances the DRG axonal response to dorsal spinal cord-derived attractants. 67LR is expressed on the DRG neuronal surface and mediates the YIGSR-induced effects. (B) Netrin-1 attracts *Xenopus* retinal growth cones in the absence of LM-111 or its YIGSR peptide. LM-111 or its YIGSR peptide converts Netrin-1-induced attraction into repulsion via β1 integrin receptors by decreasing the cAMP level in the cytosol of the retinal neuron. (C) Slit2 repels rat DRG axons in the absence of LM-111 or its YIGSR peptide. However, LM-111 or its YIGSR peptide converts this Slit2-induced repulsion into attraction by causing an increase in thecGMP level in the DRG neuron.

Subsequent to the evidence presented byHöpker et al. (1999), Chédotaland his colleagues obtained data indicating the modulation of axonal responses by laminins (Nguyen-Ba-Charvet et al., 2001). They used the axon stripe assay with rat DRG neurons to reveal the effect of LM-111 on the repulsive activity of Slit2 toward DRG axons. They also succeeded in revealing that LM-111 or the YIGSR peptide can stimulate cGMP signaling pathways in rat DRG neurons.

In order to confirm whether LM-111 or its peptide YIGSR actually play a key role in axonal turning and guidance *in vivo*, it is essential for us to analyze the formation of axonal trajectories in the *LM-111*-deficient mouse embryo. However, the *in vivo* role of LM-111 in axonal guidance is unclear, because all three different *LM-111*-deficient mice (*LAMA1*-, *LAMB1*-, and *LAMC1*-deficient mice) die early during embryogenesis (embryonic day 5-7; Alpy et al., 2005; Miner et al., 2004; Smyth et al., 1999). Therefore, it is an intriguing question as to whether LM-111, especially the LAMB1 chain, is quite crucial for the axonal response to attractive or repulsive guidance cues. An important future step to investigate the exact role of LM-111 in axonal guidance will be complete *in vivo* perturbation of LM-111or the LAMB1 chain at the time when developing axons are projecting to their targets.

In conclusion, these data suggest a mechanism for axonal guidance of a variety of neurons in which LM-111 or the YIGSR peptide plays an important role in the early formation of axonal projections toward the target. It also appears that the action of guidance cues is mainly determined by the level of cyclic nucleotidesin the neuron, which can be modulated by LM-111.

References

Adler, R; Jerdan, J; Hewitt, AT. (1985) Responses of cultured neural retinal cells to substratum-bound laminin and other extracellular matrix molecules. *Dev. Biol.*, 112, 100-114.

Alpy, F; Jivkov, I; Sorokin, L; Klein, A; Arnold, C; Huss, Y; Kedinger, M; Simon-Assmann, P;Lefebvre, O. (2005) Generation of a conditionally null allele of the Lamininα1 gene. *Genesis,* 43, 59-70.

Anton, ES; Sandrock, AW Jr; Matthew, WD. (1994) Merosin promotes neurite growth and Schwann cell migration *in vitro* and nerve regeneration *in vivo*: evidence using an antibody to merosin, ARM-1. *Dev. Biol.*, 164, 133-146.

Burgeson, RE; Chiquet, M; Deutzmann, R; Ekblom, P; Engel, J; Kleinman, H; Martin, GR; Meneguzzi, G; Paulsson, M; Sanes, J; Timpl, R; Tryggvason, K; Yamada, Y; Yurchenco, PD. (1994) A new nomenclature for the laminins. *Matrix Biol.,* 14, 209-211.

Gauczynski, S; Peyrin, JM; Haïk, S; Leucht, C; Hundt, C; Rieger, R; Krasemann, S; Deslys, JP; Dormont, D; Lasmézas, CI; Weiss, S. (2001) The 37-kDa/67-kDa laminin receptor acts as the cell-surface receptor for the cellular prion protein. *EMBO J.*, 20, 5863-5875.

Graf, J; Ogle, RC; Robey, FA; Sasaki, M; Martin, GR; Yamada, Y; Kleinman, HK. (1987) Apentapeptide from the laminin B1 chain mediates cell adhesion and binds the 67000 laminin receptor. *Biochemistry,* 26, 6896-6900.

Hamburger, V; Hamilton, HL. (1992) A series of normal stages in the development of the chick embryo. *Dev. Dyn.*, 195, 231-272.

Höpker, VH; Shewan, D; Tessier-Lavigne, M; Poo, MM;Holt, C. (1999) Growth-cone attraction to netrin-1 is converted to repulsion by laminin-1. *Nature,* 401, 69-73.

Liesi, P; Närvänen, A; Soos, J; Sariola, H; Snounou, G. (1989) Identification of a neurite outgrowth-promoting domain of laminin using synthetic peptides. *FEBS Lett.,* 244, 141-148.

Manthorpe, M; Engvall, E; Ruoslahti, E; Longo, FM; Davis, GE; Varon, S. (1983) Laminin promotes neuritic regeneration from cultured peripheral and central neurons. *J. Cell Biol.,* 97, 1882-1890.

Massia, SP; Rao, SS; Hubbell, JA. (1993) Covalently immobilized laminin peptide Tyr-Ile-Gly-Ser-Arg (YIGSR) supports cell spreading and co-localization of the 67-kilodalton laminin receptor with α-actinin and vinculin. *J. Biol. Chem.,* 268, 8053-8059.

Masuda, T; Sakuma, C; Kobayashi, K; Kikuchi, K; Soda, E; Shiga, T; Kobayashi, K; Yaginuma, H. (2009) Laminin peptide YIGSR and its receptor regulate sensory axonal response to the chemoattractive guidance cue in the chick embryo. *J. Neurosci. Res.,* 87, 353-359.

Masuda, T; Sakuma, C; Taniguchi, M; Kobayashi, K; Kobayashi, K; Shiga, T; Yaginuma, H. (2007) Guidance cues from the embryonic dorsal spinal cord chemoattract dorsal root ganglion axons. *Neuroreport*, 18, 1645-1649.

Masuda, T; Shiga T. (2005) Chemorepulsion and cell adhesion molecules in patterning initial trajectories of sensory axons. *Neurosci. Res.,* 51, 337-347.

McGarvey, ML; Baron-Van Evercooren, A; Kleinman, HK; Dubois-Dalcq, M. (1984) Synthesis and effects of basement membrane components in cultured rat Schwann cells. *Dev. Biol.,* 105, 18-28.

Ménard, S; Castronovo, V; Tagliabue, E; Sobel, ME. (1997) New insights into the metastasis-associated 67 kDlaminin receptor. *J. Cell Biochem.,* 67, 155-165.

Miner, JH; Li, C; Mudd, JL; Go, G; Sutherland, AE. (2004) Compositional and structural requirements for laminin and basement membranes during mouse embryo implantation and gastrulation. *Development,* 131, 2247-2256.

Nguyen-Ba-Charvet, KT; Brose, K; Marillat, V; Sotelo, C; Tessier-Lavigne, M; Chédotal, A. (2001) Sensory axon response to substrate-bound Slit2 is modulated by laminin and cyclic GMP. *Mol. Cell Neurosci.,* 17, 1048-1058.

Obremski, VJ; Bunge, MB.(1995) Addition of purified basal lamina molecules enables Schwann cell ensheathment of sympathetic neurites in culture. *Dev. Biol.,* 168, 124-137.

Rogers, SL; Letourneau, PC; Palm, SL; McCarthy, J; Furcht, LT. (1983) Neurite extension by peripheral and central nervous system neurons in response to substratum-bound fibronectin and laminin. *Dev. Biol.,* 98, 212-220.

Shiga, T; Kawamoto, M; Shirai, T. (2000) Longitudinal elongation of primary afferent axons in the dorsal funiculus of the chick embryo spinal cord. *Dev. Brain Res.,* 124, 25-31.

Smyth, N; Vatansever, HS; Murray, P; Meyer, M; Frie, C; Paulsson, M; Edgar, D. (1999) Absence of basement membranes after targeting the *LAMC1* gene results in embryonic lethality due to failure of endoderm differentiation. *J. Cell Biol.*, 144, 151-160.

Tessier-Lavigne, M; Goodman, CS. (1996) The molecular biology of axon guidance. *Science*, 274, 1123-1133.

Index

A

access, 119
accounting, 70, 76
acid, 33, 73, 76, 82, 122
acidic, 55, 67, 74
acylation, 73, 81, 82
AD, 65, 66, 122
adenocarcinoma, 85
adhesion(s), vii, viii, x, 30, 32, 33, 34, 35, 36, 37, 38, 39, 40, 41, 42, 43, 44, 45, 46, 47, 50, 51, 53, 54, 56, 57, 58, 61, 63, 64, 66, 67, 70, 72, 75, 81, 89, 90, 101, 112, 114, 117, 118, 119, 121, 122, 124, 128, 129, 130, 134, 135
adhesion strength, 45
adulthood, viii, 53, 58, 62
age, 91, 92
aggregation, 80
aggressiveness, 61
agonist, 95, 102
allele, 77, 78, 79, 91, 134
ALS, 60
alters, 130
amino, 31, 33, 35, 73, 74, 76, 77, 132
amino acid(s), 33, 35, 73, 74, 76, 77, 132
amnion, 42, 71
amyotrophic lateral sclerosis, 63, 65
anabolism, 102
anatomy, 66
anchorage, 38, 118
anchoring, 37, 43, 44, 47
aneurysm, 103
angiogenesis, 42, 71, 84
angiotensin receptor inhibitor, 96
ankles, 90
antibiotic, 94
antibody, 35, 113, 124, 132, 134
anti-cancer, viii, 69, 75
antigen, 43, 49
antisense, 78
apoptosis, ix, 37, 49, 55, 60, 81, 87, 92, 93, 94, 96, 97, 101, 120
appetite, 78
ARM, 134
arrests, 120
arterioles, 71
artery(ies), x, 106, 112, 117, 119
arthritis, 103
assessment, 124
astrocytes, viii, 53, 60, 61, 64, 66, 77
astrogliosis, 67
atopic dermatitis, 60, 68
ATP, 102
atrophy, 102, 103
attachment, 42, 118
attractant, xi, 131
autoantibodies, 60, 66, 116
autosomal dominant, 59, 64
autosomal recessive, 35, 88, 90, 99
awareness, vii
axonal response, vii, xi, 131, 132, 133, 135
axons, viii, x, 53, 58, 60, 63, 64, 68, 124, 131, 132, 133, 134, 135

B

bacteremia, 126, 130
bacteria, vii, viii, x, 69, 75, 78, 125, 126, 127, 128, 129, 130
bacterial pathogens, 78
basal lamina, vii, 53, 106, 107, 111, 118, 121, 128, 135
basement membrane, vii, viii, x, 29, 32, 42, 43, 44, 45, 46, 53, 57, 69, 80, 83, 84, 98, 118, 119, 120, 121, 122, 123, 124, 125, 126, 129, 130, 135
BD, 62

benefits, 103
bile, ix, 105, 106, 107, 108, 109, 111, 113, 114, 115
bile duct, ix, 105, 106, 107, 108, 109, 113, 114, 115
bioavailability, 77
biochemical, ix, 87
biological activity(ies), vii, 36, 44, 46, 48, 55, 58, 70
biological processes, 62, 119
biological roles, viii, 53
biosynthesis, 70, 74, 99
birds, vii
blastocyst, x, 117, 118, 119, 120, 121, 123
blood, 30, 54, 70, 71, 118, 130
blood vessels, 30, 54, 70
blood-brain barrier, 54
body size, 78
body weight, 96
bonds, x, 30, 31, 117, 118
bone, 33, 36
brain, 54, 58, 59, 61, 65, 66, 67, 72, 76, 78, 89, 90, 123
branching, 39, 58, 64
Brazil, 117
breakdown, 46
breast cancer, 42, 70, 83
breast carcinoma, 30, 38, 75, 83
budding, 46
bullous pemphigoid, 43

C

cachexia, 102
calcium, 60
CAM, 112
cancer, vii, viii, 29, 34, 35, 36, 38, 39, 40, 41, 44, 46, 60, 63, 65, 69, 70, 75, 80, 81, 84, 85, 102, 119
cancer cells, 34, 35, 40, 46, 60, 119
cancer progression, 60
carbohydrate(s), viii, 38, 39, 49, 55, 61, 67, 69, 75
carbon, 114
carbon tetrachloride, 114
carboxyl, 31, 45
carcinogenesis, 48, 49
carcinoma, 46, 48, 49, 67, 75, 83, 84, 130
cardiovascular disease, 126
cartilage, 59, 121
catecholamines, 95
CBP, 94
cDNA, 37, 103
cell biology, 59, 123
cell culture, 38
cell cycle, 55
cell death, 68, 83, 90, 120, 123
cell differentiation, vii, 57, 61, 119
cell division, 123
cell fate, 44
cell invasion, 37, 38
cell line(s), x, 33, 36, 39, 60, 80, 108, 111, 112, 113, 125, 126
cell signaling, 40
cell surface, vii, viii, 29, 31, 32, 34, 40, 47, 50, 54, 55, 57, 67, 69, 70, 72, 76, 77, 89, 90
cellular prion, viii, 69, 76, 81, 134
cellular signaling pathway, 41
central nervous system (CNS), 54, 57, 58, 59, 60, 61, 64, 67, 71, 75, 80, 135
cerebellum, 58, 60
challenges, 92
chicken, 66
children, ix, 87, 90
chimera, 55
CHO cells, 64
cholangiocyte, ix, 105, 107, 108, 109, 115
choline, 112
chromosome, 73, 89
circulation, x, 117, 120
cirrhosis, 114, 116
cleavage, 33, 43, 47, 48, 78, 83, 128, 130
cloning, 42
closure, x, 118, 120
clustering, 39
coatings, 41
coding, 84, 99, 109, 121
codon, 76, 78, 79
cognitive function, 61
collagen, 30, 32, 33, 34, 35, 37, 43, 44, 48, 88, 89, 96, 106, 113, 118, 119, 120, 121, 122, 130
collateral, 58
colon, 46, 48, 60, 75, 81, 102
colon cancer, 60, 81
communication, 35, 46, 89
community, 59
compensation, 123
complications, ix, 87, 90
composition, 55, 78, 106, 130
compounds, 70, 119
comprehension, 62
conduction, 92
conference, 102
constituents, 53
construction, 120
COOH, 44
coronary artery disease, 129
correlation(s), 44, 46, 60, 100, 132
cortex, 58, 59, 77, 122, 123
cortical abnormalities, 121
cortical neurons, 80

creatine, 90
Creutzfeldt-Jakob disease, 80, 82, 85
CT, 126
cues, vii, xi, 131, 132, 133, 134, 135
culture, ix, xi, 38, 41, 105, 108, 110, 112, 113, 114, 115, 131, 132, 135
culture conditions, 113
culture media, 112
cure, 98
cyst, 108, 113
cysteine, 33, 55, 64
cytokines, 60, 95
cytoplasm, 55, 73, 77
cytoplasmic tail, 35

D

defects, x, 32, 47, 59, 88, 89, 92, 113, 114, 118, 120, 121, 123
deficiency, 43, 59, 92, 93, 95, 99, 100, 101, 121
degenerate, 77
degradation, vii, x, 125, 126, 127, 128, 129, 130
demyelinating disease, 59
demyelination, 59
dendrites, 77
dendritic cell, 75, 81
dengue, 78
deposition, 33, 34, 35, 37, 44, 46, 96, 99
deprivation, 49
dermis, 32
destruction, 80, 119, 126, 128, 130
detachment, 35, 43, 94
detection, 67
developing brain, 77, 123
developmental disorder, 44
developmental process, vii, 113
diet, 112
dimerization, 73, 82
disease gene, 85
disease model, 72
disease progression, 94, 96
diseases, vii, ix, 41, 60, 72, 75, 76, 80, 85, 87, 97, 102
disorder, 35
disposition, 55
distribution, 49, 55, 59, 70, 80, 89, 122
diversity, vii, 42
DNA, 73, 95
dorsal horn, 58
Down syndrome, 61, 66
down-regulation, 42, 133
Drosophila, 64, 124
drugs, 41
dysplasia, 63

E

EAE, 60
E-cadherin, 35, 38, 110, 129
ECM, ix, 30, 34, 36, 54, 55, 56, 57, 58, 60, 61, 87, 88, 89, 90, 92, 94, 96, 106, 110, 111, 112, 113, 121
ECs, 106
editors, 122
elastin, viii, 69, 75
elbows, 90
electric field, 47
electron, 33
electron microscopy, 33
elongation, 132, 135
embryo development, vii, x, 117, 118, 120, 121
embryogenesis, 30, 120, 123, 133
embryonic stem cells, 41, 51
encephalitis, 78
encoding, 35, 42, 59, 64, 73, 85, 122
endocarditis, 130
endoderm, 106, 120, 123, 135
endothelial cells, 42, 50, 71, 78, 82, 106, 109
endothelium, 119
enlargement, 108
environment(s), 61, 75, 113
enzyme(s), 36, 39, 47, 126
ependymal, 61
epidemiology, 100
epidermis, 32, 35, 60
epidermolysis bullosa, 31, 32, 35, 37, 41, 43, 46
epithelia, 30, 70
epithelial cells, ix, 34, 38, 42, 45, 78, 103, 105, 106, 113, 115, 126, 129
epithelial layer, x, 125, 126
epithelial ovarian cancer, 84
epithelium, x, 117, 118, 119, 120, 125, 126, 129, 130
ester, 73
etiology, 76
EU, 62
eukaryotic cell, 84
evidence, vii, ix, xi, 29, 61, 82, 88, 131, 132, 133, 134
experimental autoimmune encephalomyelitis, 66
exposure, 113
extracellular matrix, vii, ix, x, 30, 33, 44, 46, 64, 65, 66, 68, 70, 77, 80, 87, 88, 98, 99, 101, 105, 106, 115, 117, 118, 119, 120, 122, 123, 124, 129, 130, 131, 132, 134
extracts, 84

F

failure to thrive, ix, 87, 90
families, 89
family members, 95
fasciculation, 58, 68
fat, 30, 78
fetus, 120
fiber(s), ix, 71, 87, 92, 94, 96
fibroblast growth factor, 63, 64, 65, 94, 101
fibroblasts, 39, 107, 108, 109
fibrogenesis, 112
fibrosis, ix, 87, 92, 93, 94, 95, 96, 97, 103, 111
filament, 47
Finland, 125, 129, 130
fluorescence, 113
Food and Drug Administration (FDA), 69, 95
force, 94, 97
forebrain, 65
formation, vii, ix, 35, 37, 39, 40, 46, 54, 58, 59, 64, 68, 76, 78, 84, 103, 105, 106, 107, 108, 110, 111, 113, 120, 121, 133, 134
fragments, 51
France, 62
functional analysis, 82
fusion, 57, 99
Fusobacterium, x, 125, 127

G

ganglion, 132, 135
gastrulation, 123, 135
gel, 113
gene expression, 60, 62, 95, 111
gene transfer, 100
genes, 35, 71, 78, 97, 98, 100, 113, 121
genetic defect, 31, 90
genetic disorders, vii, 29, 31
genetic linkage, 100
genetics, 43, 64
genotype, 44, 46, 77, 78, 79, 80, 100
germ cells, 120
germ layer, 120
gestation, x, 117, 121
gingival, vii, x, 125, 126, 128, 130
gingival epithelium, 130
gland, 113, 115, 118, 119
glia, 121
glial cells, 54, 61, 65, 77, 121
glial endfeet, 121
glioma, 61
glomerulogenesis, x, 113, 118, 121
glomerulus, x, 118, 121, 123
glutamic acid, 45
glutamine, 40
glycans, 39, 40, 58, 62, 67
glycoprotein(s), viii, 30, 40, 50, 53, 55, 60, 61, 67, 68, 69, 70, 80, 84, 89, 99, 126, 132
glycosylation, viii, 29, 38, 39, 41, 50, 68, 89, 99
gonadotropin-releasing hormone (GnRH), 63, 68
gonads, 59
grants, 62, 129
growth, x, 31, 37, 38, 45, 47, 48, 51, 54, 56, 58, 61, 65, 70, 72, 75, 94, 101, 102, 119, 125, 126, 127, 128, 131, 132, 133, 134
growth factor, 31, 37, 45, 47, 48, 56, 70, 94, 101, 102
guidance, vii, xi, 54, 58, 131, 132, 133, 134, 135

H

hair, 113, 114
hair follicle, 113
HCC, 38, 106, 112
HE, 121
healing, 35, 101
health, 129
hemidesmosome, 32, 34, 35, 37, 38, 48
hepatic failure, 114
hepatic injury, 112
hepatitis, 116
hepatocellular carcinoma, 38, 49, 83, 106, 114
hepatocytes, ix, x, 105, 106, 107, 110, 111, 114, 115
hippocampus, 65, 71
histogenesis, 122
histones, 73
homeostasis, 97
hormone(s), x, 49, 63, 64, 66, 117, 120
host, ix, 78, 87, 126, 130
human brain, 78, 82
human immunodeficiency virus (HIV), 78
human skin, 42
hybrid, 76
hypertrophy, 94, 101, 102
hypogonadism, 59
hypoplasia, 59
hypoxia, 60, 65

I

iatrogenic, 76
ideal, 96, 97
identification, 42, 130
IFN, 95

immigration, 59
immune system, x, 102, 117, 119
immunity, 68
immunoglobulin, 38
immunoprecipitation, 126
immunoreactivity, 60, 68
improvements, 95, 96
in vitro, ix, 37, 51, 105, 108, 109, 110, 111, 112, 113, 114, 116, 119, 124, 128, 132, 134
in vivo, vii, 30, 101, 102, 108, 111, 112, 113, 114, 122, 132, 133, 134
incidence, 76, 90
incubation period, 81, 82, 83
incubation time, 76, 77, 83, 84
induction, 34, 41, 65, 81, 109, 114
infection, ix, 76, 77, 84, 105
infertility, 59
inflammation, ix, 59, 87, 92, 93, 94, 95, 96, 97
inflammatory cells, 95
inflammatory disease, 60
inherited disorder, 88
inhibition, 49, 57, 75, 94, 96, 97, 103, 112
inhibitor, 58, 63, 95, 97
injury(ies), 36, 61, 62, 66, 67, 94, 101, 112, 114
inositol, 76, 94
insertion, 43
insomnia, 76
insulin, 82, 94, 101, 102
integration, 36
integrin(s), vi, viii, x, 29, 31, 32, 33, 34, 35, 36, 37, 38, 39, 40, 41, 43, 44, 45, 46, 47, 48, 49, 50, 54, 56, 57, 58, 64, 69, 70, 72, 73, 80, 88, 90, 93, 94, 98, 99, 108, 110, 111, 112, 113, 114, 115, 117, 119, 120, 121, 122, 128, 132, 133
integrity, viii, x, 34, 58, 69, 90, 94, 100, 118, 121, 125, 126
intercellular contacts, 61
interference, 36, 71
internalization, 78
interneurons, 57, 80
interstitial lung disease, 35, 39
intervention, 102
intestine, 106
invertebrates, 57
ions, 106
ischemia, 61, 65, 67
isolation, 115
Italy, 90, 100
IV collagenase, 123

J

Japan, 29, 41, 131
joint destruction, 103

K

keratin, 32
keratinocyte(s), 32, 34, 35, 36, 37, 40, 43, 44, 45, 46, 47, 48, 49, 50, 60, 126, 129
kidney(s), 38, 45, 71, 96, 103, 113, 114
knees, 90
Korea, 69

L

laceration, 101
lactobacillus, 126, 130
lactose, 64
lamination, 54, 67, 122, 123
laminin heterotrimer, 30
laminin-5, vii, 29, 30, 34, 43, 44, 45, 46, 47, 48, 50, 51, 70, 126
lateral sclerosis, 60
lattices, 50
lead, viii, ix, 30, 35, 76, 77, 88, 89, 90, 92, 94, 97, 133
learning, 61
lesions, 59, 60, 62, 63, 65
life cycle, 77, 84
ligand, 38, 42, 45, 68
light, 113
liver, vii, ix, 75, 105, 106, 107, 108, 109, 110, 111, 113, 114, 115, 116
Liver progenitor, ix, 105, 114, 115
localization, 34, 40, 114, 134
loci, 82, 84
locus, 100
lumen, ix, 105, 107, 108, 110, 113, 118
lung cancer, 84
lung disease, 50
Luo, 62
lysosome, 57, 76

M

machinery, 97
macromolecules, 54
macrophages, 60, 61, 65, 71, 75, 95, 102
magnetic resonance image, 63
malignancy, 62
malignant cells, 60, 124
mammalian brain, 121
mammalian cells, 76, 78, 85
mammals, vii, viii, 30, 34, 58, 69, 70, 73

man, 115
mapping, 82
Marfan syndrome, 96, 103
mass, 95, 96, 128
materials, 70, 127
matrix, ix, xi, 33, 34, 35, 36, 44, 47, 48, 64, 67, 68, 71, 89, 96, 106, 115, 119, 126, 129, 130, 131
matrix metalloproteinase, 36, 48, 106, 126
MB, 135
media, x, 125, 126, 127, 128
median, 63
medical, 41
MEK, 94
melanoma, 39, 46, 49, 50, 61, 63
membranes, vii, x, 42, 89, 117, 118
memory, 61, 67
memory formation, 61, 67
mesangial cells, 114
mesenchymal stem cells, 50
mesenchyme, 106
mesothelium, 106
metalloproteinase, 36, 47
metastasis, viii, 39, 60, 61, 69, 70, 72, 75, 83, 135
mice, 31, 35, 37, 42, 43, 49, 58, 61, 66, 67, 76, 77, 80, 81, 82, 83, 84, 85, 89, 91, 92, 93, 94, 95, 96, 97, 98, 99, 100, 101, 102, 103, 106, 107, 109, 112, 113, 114, 115, 118, 119, 120, 123, 133
microscope, 113
migration, vii, viii, 29, 30, 33, 34, 35, 36, 37, 38, 39, 40, 41, 42, 43, 44, 46, 47, 48, 49, 50, 53, 54, 56, 57, 59, 61, 63, 64, 65, 67, 68, 70, 72, 75, 81, 89, 111, 115, 118, 119, 120, 121, 129, 132, 134
Ministry of Education, 41
mitogen, 47
MMP, 36, 47, 106
modelling, 67
models, ix, 87, 91, 94, 96, 97, 100, 112, 113
modifications, viii, 29, 133
modules, 31
molecular biology, 135
molecular mass, 30
molecular weight, 38, 47, 70, 127
molecules, vii, viii, x, 33, 34, 35, 37, 39, 40, 41, 53, 54, 55, 56, 59, 61, 63, 64, 66, 68, 69, 70, 72, 89, 97, 101, 117, 118, 122, 124, 126, 131, 134, 135
monoclonal antibody, 49
monomers, 55
morphogenesis, vii, ix, 61, 70, 72, 80, 98, 105, 106, 107, 108, 109, 113, 114, 115, 120, 124
morphology, 34, 45, 103, 113
morula, 118
mosquitoes, 78
motif, 39, 48, 81
mRNA, 55, 83, 84, 102, 122
mucous membrane(s), 35
multiple sclerosis, 59, 60, 63, 64, 65, 67, 68
muscle atrophy, 95, 101, 102, 103
muscle contraction, 101
muscle mass, 94, 96
muscle stem cells, 101
muscles, 92, 95, 96, 102
muscular dystrophy(ies), ix, 85, 87, 88, 89, 90, 91, 98, 99, 100, 101, 102, 103, 121, 123
mutant, 39, 58, 83, 91, 98, 100, 101, 109, 113, 123
mutation(s), ix, 34, 35, 37, 39, 43, 44, 45, 46, 57, 59, 64, 67, 87, 89, 90, 91, 97, 98, 100, 120
myelin, 58, 67
myogenesis, 99, 101, 102
myopathy, 89, 99

N

NaCl, 126
necrosis, ix, 87
nematodes, vii
nephrotic syndrome, 35, 39, 50
nerve, x, 30, 58, 60, 61, 65, 68, 71, 85, 93, 94, 98, 118, 121, 134
nerve growth factor, 94
nervous system, vii, viii, 53, 54, 58, 60, 64, 65, 67, 68, 71, 75, 80, 135
neuroblasts, 57, 64
neurodegeneration, 60, 72, 80
neurodegenerative disorders, 75
neurogenesis, 61, 65, 121
neuroinflammation, 60
neurological disease, viii, 53, 63, 66
neuronal cells, 78, 82, 83
neurons, viii, 53, 57, 58, 59, 63, 66, 68, 77, 121, 132, 133, 134, 135
neuropathic pain, 61
neuroprotection, 60, 65
neurorepair, 62
neutrophils, 48
NH2, 42
nonsense mutation, 43
normal development, 45
nuclei, 38, 94
nucleotides, xi, 131, 132
nucleus, 73, 77
null, 35, 37, 38, 39, 43, 58, 91, 100, 101, 120, 123, 134
nutrients, x, 117, 119

O

obesity, viii, 69, 75, 81
OH, 49
oleic acid, 73
olfactory nerve, 59
oligodendrocytes, viii, 53
oligodendroglia, 62
oligomers, 81
oligosaccharide, 49
oocyte, 118
optic nerve, 63
organ, 114, 120
organelles, 97
organs, vii, 76, 112, 118
ovarian cancer, 82
oxidative stress, 78, 80

P

p53, 94
pancreas, 76
paralysis, 91, 92
parenchymal cell, 106
parvalbumin, 80
pathogenesis, x, 60, 64, 76, 92, 99, 100, 106
pathogens, x, 125, 126, 130
pathologist, 59
pathology, vii, viii, ix, 53, 62, 87, 89, 90, 91, 93, 94, 95, 96, 97, 101, 103
pathophysiological, ix, 88
pathophysiology, 60
pathways, 49, 57, 90, 97, 102, 109, 132
PCR, 62, 108
peptide(s), xi, 48, 55, 70, 73, 74, 75, 82, 131, 132, 133, 134
perfusion, x, 117
peri-implantation stage, 120
periodontal, x, 125, 126, 128, 129
periodontitis, 129
peripheral nervous system, 71
peripheral neuropathy, 91, 100
pH, x, 125, 126, 127, 128
pharmaceutical, 94
phenotype(s), vii, 31, 35, 44, 46, 90, 91, 97, 100, 120
Philadelphia, 122
phosphate, 94
phosphorylation, 37, 38, 41, 44, 49, 102
physicians, vii, 29
physiology, 58, 61, 122
PI3K, viii, 29, 36, 37, 38, 40, 41, 90, 99
PI3K/AKT, 99
placenta, x, 71, 117, 119, 120, 122
plaque, 64
plasma membrane, 71
plasminogen, 46, 47, 56, 65
platform, 40, 41
playing, 54, 61
PM, 65, 67
point mutation, 91, 100
polar, 123
polarity, 108, 113, 115, 116
polarization, 120, 123
polymerization, 30, 31, 33
polymorphism(s), 76, 77, 78, 79, 85
polypeptide, x, 117, 118
population, 60, 82, 90, 100
portal vein, 106, 107, 109, 110, 111
potential benefits, ix, 45, 87
precursor cells, 59
pregnancy, vii, x, 117, 118, 119, 120, 121, 124
premature death, ix, 87
prevention, 85
prions, 81, 84
probiotic, 129, 130
progenitor cells, ix, x, 58, 65, 67, 105, 106, 107, 110, 111, 112, 114, 115
prognosis, 38, 75, 84, 112
proliferation, x, 30, 37, 38, 41, 42, 49, 50, 58, 65, 67, 71, 75, 81, 89, 94, 96, 106, 110, 117, 118
propagation, viii, 69, 76, 82, 84
prostate cancer, 48, 85, 128, 130
prostate carcinoma, 48
protease inhibitors, 55
proteasome, 97
protection, x, 117, 119, 120
protective role, 78, 80
protein kinase C, 36
protein kinases, 47
protein misfolding, 72
protein sequence, 55
protein synthesis, 94
proteinase, 129, 130
proteins, vii, viii, ix, 34, 39, 40, 41, 45, 53, 54, 55, 57, 58, 69, 70, 72, 73, 76, 77, 87, 88, 89, 90, 92, 96, 97, 105, 106, 111, 112, 113, 120, 122, 130
proteoglycans, 34, 36, 54, 57, 70, 74, 111
proteolysis, 33, 36
prototype, 55
pseudogene, 64
puberty, 59
purification, 50
P-value, 79

R

RE, 134
reactive oxygen, 78
reading, 73
receptors, vii, viii, 29, 31, 34, 36, 37, 41, 45, 48, 68, 70, 71, 72, 83, 98, 122, 133
recognition, 39, 55, 59
recombinant proteins, 41
recombination, 91
recovery, ix, 61, 64, 65, 67, 88, 101
regenerate, 94
regeneration, ix, 58, 61, 62, 65, 87, 90, 92, 93, 94, 101, 103, 106, 111, 115, 134
regenerative capacity, 91, 97, 99, 100
rejection, x, 117, 119
relevance, viii, 53, 66, 115
remyelination, 59, 63, 64, 65
repair, 42, 112
repellent, xi, 131
replication, 111
reproduction, 118
repulsion, 133, 134
requirements, 123, 135
researchers, vii, 29
residues, 33, 37, 47, 48, 70, 74
resistance, 30, 75, 76, 78
response, vii, xi, 60, 77, 83, 92, 94, 110, 114, 131, 132, 133, 135
reticulum, 89
retrovirus, 109
ribosome, 73
RNAs, 83
rodents, 76, 111, 121
root, 132, 135

S

scatter(ing), 35, 39, 45, 46
sclerosis, 64
SDS-PAGE, x, 125, 127
secrete, 36, 61, 106, 120
secretion, 36, 38, 58, 59, 60, 119
senescence, 81
sensing, 81
sensitivity, 66
sensory axons, xi, 131, 135
serine, 37, 48, 128
serum, ix, 66, 95, 105
sex, 77
shape, 70
sheep, 76, 83
showing, xi, 61, 102, 120, 131
sialic acid, 59
signal transduction, 40, 92
signaling pathway, viii, 29, 34, 37, 38, 39, 49, 85, 90, 93, 98, 107, 109, 133
signalling, vii, 48, 57, 59, 61
signals, vii, 106, 107, 108, 109, 110, 113, 115, 132
signs, 60, 62, 66, 95
skeletal muscle, vii, x, 30, 57, 71, 88, 89, 93, 94, 98, 99, 100, 102, 103, 118, 121
skin, vii, 29, 30, 31, 32, 34, 35, 37, 42, 44, 45, 47, 60, 70, 71, 113
skin cancer, 42
smooth muscle, 71, 118
SNP, 76
South Korea, 80
SP, 129, 134
Spain, 53, 62
species, 76, 78, 126, 128, 130, 132
spinal cord, xi, 58, 61, 64, 65, 131, 132, 133, 135
spinal cord injury, 61, 64, 65
spleen, 75, 76
squamous cell carcinoma, 37
SS, 134
stability, 30, 70, 92, 101, 128
stabilization, 34, 45, 61, 119
standard of living, 98
state, 40, 57, 99
stem cells, 41, 51, 67, 115
stimulation, 37, 45, 82
stress, 76, 84, 99
stroke, 65
stroma, x, 117, 118, 119
structural changes, 129
structure, vii, ix, 30, 33, 34, 38, 42, 54, 55, 70, 82, 83, 84, 85, 97, 105, 108, 110, 113, 126, 128, 130
structure formation, ix, 105
subdomains, 31
substitution, 76
substrate(s), 35, 41, 48, 57, 118, 128, 130, 135
sulfate, x, 34, 36, 44, 64, 70, 74, 76, 111, 117, 118
Sun, 64, 65, 80
suppression, 37, 60, 71
surveillance, 85
survival, vii, viii, 42, 53, 54, 58, 77, 90, 94, 99, 101, 115, 121
susceptibility, 60, 68, 78, 130
symptoms, 59, 60, 93, 97, 98, 101
synapse, 54
synaptic plasticity, 58
syndrome, 35, 55, 59, 62, 63, 64, 65, 66, 67, 68, 76
synthesis, 96, 122

T

tamoxifen, 112
target, ix, 58, 80, 85, 88, 95, 98, 113, 134
techniques, 39
technology, 97, 98
teeth, 130
testis, 71
TGF, 60, 71, 96, 103, 107, 109
therapeutic approaches, 89
therapeutic interventions, 98
therapeutic targets, 97
therapy, ix, 41, 80, 85, 87, 90, 93, 94, 96, 97, 98, 100, 101, 103
threonine, 37, 48
thyroid, 75, 83
tibialis anterior, 92, 95, 96
tissue, vii, ix, 30, 31, 36, 38, 47, 55, 57, 58, 68, 70, 72, 77, 80, 89, 92, 93, 96, 98, 103, 105, 111, 113, 115, 121, 123, 126, 128
tissue homeostasis, vii
TLR4, 81
TNF, 71, 95
TNF-α, 71
tooth, 129
toxicity, 82
toxin, 45
transcription, 107, 114, 115
transcription factors, 107
transducer, 114
transforming growth factor, 49, 101
transgene, 95
translocation, 38, 73
transmembrane glycoprotein, 56
transmission, 76
treatment, ix, 38, 61, 87, 92, 93, 94, 96, 97, 99, 102
trial, 94, 95
trophoblasts, x, 117
tryptophan, 114
tumor(s), viii, 37, 38, 41, 46, 48, 49, 61, 69, 70, 71, 72, 75, 76, 83, 84, 102, 112, 119
tumor cells, 119
tumor development, 49
tumor growth, 46, 70, 72, 75
tumor invasion, 49
tumor necrosis factor, 71
tumor progression, 37, 38, 48, 84
tumorigenesis, 37, 60
tumours, 60
turnover, 106
tyrosine, 38, 40, 49, 99

U

ubiquitin, 97, 102, 103
ubiquitin-proteasome system, 97
underlying mechanisms, 98
urokinase, 46, 56, 65
USA, 67, 82, 83, 102, 126
uterine remodeling, vii, x, 118, 121
uterus, x, 71, 117, 118, 121, 122

V

vascularization, 113
vasculature, 82
vector, 109
vessels, 30, 70, 89, 119
viral vectors, 85
viruses, viii, 69, 70, 75, 78
VLA, 47

W

waste, 102
water, 106
weakness, 88, 90
weight gain, 96
white blood cells, 71
white matter, 60, 90, 100
wild type, 40
World Health Organization (WHO), 85
wound healing, vii, 29, 33, 35, 36, 39

X

xenografts, 82

Y

Y chromosome, 64
yeast, 76
yield, 67, 114
yolk, 120
young adults, 59